Fungi in biological
control systems

To Jackie, Gareth, Stuart and Martin

Fungi in biological control systems

Edited by M.N. Burge

Manchester University Press
Manchester and New York
Distributed exclusively in the USA and Canada
by **St. Martin's Press**

Published by Manchester University Press
Oxford Road, Manchester M13 9PL, UK
and Room 400, 175 Fifth Avenue, New York, NY 10010, USA

Distributed exclusively in the USA and Canada by
St. Martin's Press, 175 Fifth Avenue, New York,
NY 10010, USA

British Library cataloguing in publication data
Fungi in biological control systems.
   1. Crops. Pests. Biological control.
   Use of fungi
   I. Burge, M.N.
   632'.96

Library of Congress cataloging in publication data
Fungi in biological control systems/edited by M.N. Burge.
     p. cm.
   ISBN 0-7190-1979-: $40.00 (U.S.: est.)
   1. Agricultural pests—Biological control 2. Fungi in
agriculture. I. Burge, M.N.
  SB975.F86 1988
632'.96—dc19

ISBN 0-7190-1979-6 *hardback*

Typeset in Hong Kong
by Best-set Typesetter Ltd

Printed in Great Britain
by Biddles Ltd, Guildford and King's Lynn

Contents

Preface

The number of textbooks on biological control that have appeared in recent years reflects a strong wave of interest in the development of economical pest control methods which promise to reduce the intensity of chemical pesticide usage. Most of these texts explore the general principles and practices of the subject and a few have concentrated on microorganisms as biological control agents. Among the microorganisms, fungi and bacteria in particular exert controlling influences on other organisms in nature which, coincidentally, may be harmful, directly or indirectly to man. It is clear that, in harnessing organisms as biological control agents, fungi and bacteria have a major advantage in that their propagules can be raised in huge numbers in artificial media to produce suspensions which can be applied in the manner of conventional (chemical) pesticides. Compared with fungi, however, the number of bacteria under investigation is relatively small, although a few have great potential. Most of the bacteria under investigation are being considered as agents of insect control; *Bacillus thuringiensis* is already well established as a successful biological control agent against a range of serious insect pests. In contrast, there are many different species or strains of fungi in laboratories throughout the world with established or potential value for the control of insects, mites, weeds, eelworms and plant pathogens.

The text is therefore a compilation of chapters by active researchers from the UK, Australia and the USA, reviewing the major areas of biological control research in which fungi are either agents or targets. Following my introductory consideration of the scope of fungi in biological control, the arrangement of chapters is generally based on the target pest, in the order of insects (horticultural and agricultural), weeds, pathogens and eelworms. Since several biological control products are commercially available, a chapter is included on the techniques of mass production of entomogenous fungi. A chapter is also included on the potential for genetic manipulation of candidate fungi which may lack virulence or other desirable qualities for field success. The final chapter exemplifies what may be done with fungal biological control agents as components of integrated pest management. Although this last chapter is concerned only with the control of plant pathogens, the principles are, of course, applicable to other pests.

It is hoped that the book will be informative, not only to researchers in the many aspects of biological control involving fungi, but also to researchers and advanced students of mycology, entomology, agronomy, weed science and nematology. Additionally, the text should be of interest to pesticide scientists and to persons concerned about the impact of pesticides on the environment.

On behalf of all the contributors, I thank those colleagues who have assisted with their critical readings and comments on particular chapters. Special thanks are afforded to Denis Burges for criticisms of several chapters and for his encouragement and advice.

M.N. Burge

Contributors

E.B. Adams, US Department of Agriculture, Washington State University, N222 Havana, Washington D.C. 99202, USA.

M.C. Bartlett, Chemical & Agricultural Products Division, Abbott Laboratories, North Chicago, IL 60064, USA.

J.P. Blakeman, Department of Mycology and Plant Pathology, The Queen's University of Belfast, Newforge Lane, Belfast BT9 5PX, UK.

M.N. Burge, Biology Division, Todd Centre, Department of Bioscience and Biotechnology, University of Strathclyde, Glasgow G4 0NR, UK.

R. Charudattan, Plant Pathology Department, University of Florida, Gainesville, FL 32611, USA.

R.C. Cooke, Department of Botany, University of Sheffield, Sheffield S10 2TN, UK.

J.L. Faull, Biology Department, Birkbeck College, Malet Street, London WC1E 7HX, UK.

A.T. Gillespie, Institute of Horticultural Research, Worthing Road, Littlehampton, Sussex BN17 6LP, UK.

J.B. Heale, Department of Biology, King's College London, Kensington Campus, Campden Hill Road, Kensington, London W8 7AH, UK.

S.T. Jaronski, Chemical & Agricultural Products Division, Abbott Laboratories, North Chicago, IL 60064, USA.

J.A. Lewis, Soilborne Diseases Laboratory, Plant Protection Institute, Agricultural Research Service, US Department of Agriculture, Beltsville, MD 20705, USA

K. Lewis, Department of Botany, University of Sheffield, Sheffield S10 2TN UK.

G.C. Papavizas, Biocontrol of Plant Diseases Laboratory, Plant Sciences Institute, Agricultural Research Service, US Department of Agriculture, Beltsville, MD 20705, USA.

R.J. Quinlan, Applied Power Technology, Brompton Road, London, UK.
G.R. Stirling, Plant Pathology Branch, Queensland Department of Primary Industries, Meiers road, Indooroopilly, Queensland 4068, Australia.
J.M. Whipps, Department of Microbiology and Plant Pathology, Glasshouse Crops Research Institute, Littlehampton, West Sussex BN17 6LP, UK.

1 *M.N. Burge*

The scope of fungi in biological control

1.1 Introduction

In recent years the stature of biological control as a viable practice in modern agriculture and horticulture has increased dramatically. This is clearly reflected in the steadily increasing number of symposia and publications including reviews and textbooks and, most significantly, in the number of biocontrol products now commercially available that can be successfully employed in pest management programmes. There are several reasons for this accelerated interest in biological control, not least of which is a general increase in public awareness of the potential ecological hazards posed by the use of pesticides. Less well known to the public, but equally alarming, is the fact that many of these chemicals have had their effectiveness greatly reduced as a result of the emergence of resistance in pest populations. This phenomenon is also very costly, since the resulting loss of revenue to pesticidal companies is considerable. The problem arises from the fact that many pesticides have such specific sites of attack on the cellular biology of the pest that a single mutation in a pest genome has been sufficient to alter or remove a sensitive site. It is much less likely that pests would develop resistance to biological control agents because the genetic changes in the pest necessary to achieve resistance would ostensibly be more substantial, except perhaps where a host-specific toxin is involved. Alternative methods of pest control, such as the development of resistant plant cultivars, have often been too slow, and economic pressure on land use has limited some of the traditional cultural techniques of control such as crop rotation.

All major pests have natural enemies such as predators, parasites and competitive antagonists, and successful biological control depends on the manipulation of these enemies by various means to facilitate the reduction of pest populations below the economic threshold. Total eradication is seldom feasible. The two basic methods of biological control are the 'classical' and the 'inundative'. In the former an exotic beneficial (controlling) organism is introduced into an area where an exotic pest has become established in the absence of the natural enemies that existed in its original home. Following systematic inoculations of control agent into the pest population the agent may have the ability to progress to epidemic proportions. Examples of the use of introduced fungi to control exotic weeds are provided by Adams (Chapter 6). 'Inundative' methods involve single or multiple applications of an agent — indigenous or introduced — in much higher concentrations of propagules than normally encountered by the pest. Terms such as 'mycoherbicides' and 'mycoinsecticides' have consequently been coined for fungi formulated for use in this way. Charudattan (Chapter 5) also discusses 'inoculative' (or augmentative) methods in which an organism is introduced and manipulated to enhance its establishment and control capabilities. The inundative use of indigenous species promises to be the most useful method of control using fungi, and several examples of this are described elsewhere in this text.

There are many categories of biological control agent operating against each major type of pest. Predatory, phytophagous and parasitic insects are probably the major agents used in biological control programmes to date, but a rapidly increasing number of parasitic and pathogenic nematodes, protozoa, bacteria, rickettsiae, viruses and fungi are under investigation as potential weapons of biological control. The involvement of microorganisms in biological control has been the major content of texts by Burges (1981), Charudattan & Walker (1982), Kurstak (1982), Cook & Baker (1983), Deacon (1983) and Windels & Lindow (1985). Fungi and bacteria have distinct advantages, not least of which is the ease with which they can be cultured to produce the necessary inundative concentrations of inoculum. The propagules are usually small enough to be applied using standard spraying techniques to hit the target pest, or to be brought into the pest's ecosphere in the manner of a conventional chemical pesticide. Pathogenic microorganisms have the added potential advantage of the capacity for rapid reproduction in or on the host pest, and further dissemination in the habitat. Depending on the individual situation, the inoculum may or may not require repeated application, but in an ideal situation the biocide would sustain itself at a level sufficient to hold the pest below the economic threshold for the duration of a crop. The performance of a biological control agent is even more sensitive to fluctuating climatic conditions than conventional pesticides; consequently greater consideration must be given to the timing of application of a biological agent to ensure that the latter becomes established.

1.2 Fungi as biological control agents

Fungi have evolved many mechanisms for extracting food materials from a wide variety of habitats. During this evolution, the problem of competition for food and/or space posed by other organisms in a fungal habitat has resulted in development by certain fungi of methods of antagonism or suppression of these competitors. The food source of a fungus, or organism with which it competes successfully for food or space, may be a living organism harmful to man either directly or indirectly. It is such fungi that have potential as agents of biological control.

The major pests confronting man are insects, weeds, and plant pathogens. Insects ravage crops and transmit disease organisms of man, his stock and his crops. Weeds compete with crops and choke and poison waterways, as well as harbouring harmful pests and pathogens. Plant diseases caused by fungi, bacteria, viruses, mycoplasmas, nematodes and higher plants lead to severe losses as a result of the continuous impairment of crop physiology. General texts on pests and diseases have been written by Fox Wilson (1960), Gram *et al.* (1969), Stapely & Gayner (1969), Fletcher (1974), Jones & Jones (1974) and Kranz *et al.* (1979).

All of these pests and disease organisms, with the possible exceptions of viruses, bacteria and mycoplasmas, may be subject to a degree of natural control by fungi, which are either predatory, parasitic or antagonistic. Thus, these fungi and their attributes are the subjects of investigations in our quests for successful biological control agents.

1.2.1 *Predation*

Predaceous fungi feed on protozoa and nematodes in soil, dung, rotting wood and mosses. They are characterised by the development of specialised traps with which they capture their prey before feeding on them after death. Predaceous fungi occur in diverse taxonomic groups; they entirely constitute the order Zoopagales of the Zygomycetes and also are numerous in the family Moniliaceae of the Fungi Imperfecti. The Zoopagales are obligate predators and feed predominantly on *Amoeba* and other protozoa, although a few species are large enough to trap nematodes. Their taxonomy has been reviewed by Duddington (1973). Predaceous members of the Moniliaceae feed almost exclusively on nematodes but are not obligate and may exist saprophytically in the absence of prey. Various mechanisms of prey capture are employed by different species of predaceous Moniliaceae. Species such as *Arthrobotrys oligospora* produce a sticky network of hyphae to entangle the nematodes. Others produce adhesive knobs on the ends of lateral branches (e.g. *Dactylella ellipsospora*), and some such as *D. lysipaga* produce rings of cells in the form of a noose which snares the prey nematode. In some species (e.g. *A. dactyloides*) the rings of cells rapidly enlarge on contact with the prey to constrict and strangle it. Details of mechanisms whereby nematodes trap eelworms are provided by Duddington (1957). Little is known about the biochemistry and physiology of the nematode traps, but they develop only in the presence of the nematode prey (Iffland & Allison, 1964; Higgins & Pramer, 1967). The potential of predaceous as well as parasitic fungi in the control of nematodes is discussed by Stirling (Chapter 10).

1.2.2 *Parasitism*

Parasitic fungi invade the living body of a host alga, fungus, higher plant or animal. Such fungi feed and reproduce at the host's expense while not contributing to its welfare. Some cause no noticeable damage to the host. The majority cause physiological and/or physical damage: these are called pathogens. They impair the host's health, i.e. they cause a disease marked by detectable changes in function or morphology (symptoms) which often appear in regular sequence, collectively referred to as the disease syndrome.

Some fungi are predominantly saprophytic but may become parasitic under favourable conditions. These are the facultative parasites. Generally they are not highly specialised as parasites but are often highly virulent, causing rapid death of their hosts. In contrast, some fungi are obligate parasites and grow exclusively on living tissues. These fungi are highly specialised as parasites and establish delicately balanced physiological relationships with their hosts. A host may show only mild symptoms which do not reduce its lifespan, but in certain host–parasite associations such as in crop monocultures or in dense stands of genetically uniform, perennial weeds damage can be considerable.

In terms of biological control agents the targets of pathogenic fungi are to be found among weeds, insects, nematodes and other fungi.

1.2.2.1 *Weed pathogens.* There are more than 8000 species of parasitic fungi on plants. This is far greater than the combined number of all other plant parasites (bacteria, nematodes, viruses, mycoplasmas and higher plants). A thorough analysis of the many types of interaction between parasitic fungi and their host plants has been presented by Luttrell (1974). All plant species are susceptible to attack by fungi: plant pathogens occur in all major groups of fungi from the slime moulds such as *Plasmodiophora brassicae*, which causes club root of brassicas, to the complex agarics such as the honey fungus *Armillaria mellea* which causes root rots of standing trees and shrubs. Some facultative pathogenic fungi have an extremely wide host range. *Botrytis cinerea*, for example, may attack as many as 250 different hosts covering a wide taxonomic spectrum. In contrast, many pathogenic fungi are restricted to closely related plants, e.g. *Venturia inaequalis* causes scab of fruits and foliage of plants belonging only to the genus *Malus*. Furthermore, some obligate parasitic fungi are so specialised that they have developed an intimate relationship with only one host species, as in the order Erysiphales (Ascomycetes), causal agents of powdery mildews. Specialisation is taken to its ultimate by some races of plant parasites which can attack only certain races or lines of a single host species. Thus, there are about 250 different races of *Puccinia graminis tritici*, causal agent of black stem rust of wheat, which vary in their pathogenicity to different cultivars of the host. In terms of potential biological control agents, most of the inundative techniques using indigenous fungi involve the facultative pathogens with a very narrow host range, whereas some of the most successful classical methods have employed highly specific obligate pathogens such as the rusts and powdery mildews.

Any surface tissue of a host is vulnerable to inoculation and penetration. Entry into a plant can be through wounds, through natural openings (stomata, hydathodes, lenticels, nectaries) or by direct penetration in which the fungus must induce physical and chemical degradation of resistant barriers which — depending on the nature of the plant organ under attack — can include wax, cutin, cellulose, pectin, lignin, suberin, etc. Because of the progressive build-up of such resistant chemicals, older plants are generally less susceptible to attack through direct penetration than young plants. Thus it is obvious that due attention must be paid to the effect of plant age on plant susceptibility when determining the optimum time for application of a fungal biocontrol agent. In addition, it may also be necessary to consider the optimum time of day for application; for example, some plants are vulnerable to penetration only when the stomata are fully open, which may be for a relatively short period. Inoculum applied outside this time may perish, as a result of desiccation or the effects of ultraviolet irradiation, or may be leached away by rain.

The parasitic fungus may or may not be able to live within the host after successful penetration. The parasite must, of course, gain access to its food supply: plants have evolved numerous ways of preventing this. Generally the host plant presents the fungus with an unfavourable environment for its further development by the rapid production of a wide variety of physical and chemical impediments (Albersheim & Anderson-Prouty, 1975). An example of the problem presented by a host's resistance mechanism to the potential development of a mycoherbicide is provided by our work with bracken (*Pteridium aquilinum*). Bracken is a serious invasive weed of upland pasture and of woodland (Taylor, 1986). The cost of control by herbicides, such as asulam, is often prohibitive in relation to the value of the land. Since the weed forms extensive canopies of more or less gentically uniform stands, it is a very suitable target for a biological control agent (Lawton, 1986). Bracken has very few serious natural enemies, but it is susceptible to a small number of fungal diseases, including that known as curl-tip caused by *Phoma aquilina* and *Ascochyta pteridis*. In our studies on the potential use of these fungi to control bracken (Burge & Irvine, 1985; Burge *et al.*, 1986; Irvine *et al.*, 1987) we have found that the major impediment to successful field trials is the rapid formation of post-infectional physical barriers by the host in the path of the invading hyphae (McElwee, 1987). Although many penetration events can be established in bracken in the field, the rapid formation of lignified ingrowths (papillae) of the anticlinal epidermal walls prevents further tissue invasion by the fungi. In some areas, however, the bracken does become naturally diseased with the curl-tip pathogens, and McElwee (1987) has shown that such bracken is significantly less able to produce lignified papillae.

In an effective host–pathogen association the fungus overcomes the inherent resistance mechanisms. Continued invasion of the plant is facilitated by the production of enzymes and often toxins and hormones. The potential use of fungal toxins as herbicides has received scant attention, but it may be an attractive prospect for biological weed control. Walker & Templeton (1978) demonstrated a degree of selective activity of culture filtrates of *Colletotrichum gloeosporioides* f. sp. *aeschynomene* to northern joint-vetch (*Aeschynomene virginica*).

The control of weeds with plant pathogens has received considerable attention in the literature and is the subject of the book by Charudattan & Walker (1982). Several mycoherbicidal formulations are available commercially (Templeton & Greaves, 1985). The role of fungi in weed control is reviewed in the present work by Charudattan (Chapter 5) and by Adams (Chapter 6).

1.2.2.2 *Insect parasites*. A large number of fungi are parasitic on insects and small arthropods such as mites and spiders. Insect-parasitic fungi are called entomopathogens. They include some chytrids, almost all members of the Entomophthorales, a few yeasts, numerous Ascomycetes and Fungi Imperfecti, and the Basidiomycete *Uridinella*. Reviews on the identification and use of entomopathogenic fungi are included in the texts edited by Burges & Hussey (1971) and Burges (1981). Like plant pathogenic fungi, the entomopathogens can attack an insect by penetrating directly through the protective barriers of the body wall. The insect's integument may be degraded partly physically and partly enzymically. There is much variation between strains of a fungal species, a pertinent subject for research. Generally, after penetration, yeast-like hyphal bodies multiply in the haemocoel producing toxic metabolites which kill the host. A more extensive mycelial phase then develops invading virtually all the host organs, culminating in penetration of the cuticle and production of infective units on the exterior of the host. Relatively little is known about the insect's resistance mechanisms to fungi but cellular and humoural defence reactions exist. Insect responses to microbial infections have been reviewed by Boman (1981).

Because toxins play an important role in the attack of insects by some fungi, there has been interest in their potential development as insecticides, again varying with fungal strain. Generally, however, these toxins are of low potency and many kill non-target hosts. Most of these toxins are gut poisons or are produced within the haemocoel. An extensive search may be rewarding; momentum is provided by the success of the endotoxins produced by the entomopathogenic bacterium *Bacillus thuringiensis* (Fast, 1981) which is commercially available world wide in a variety of forms for the control of Lepidoptera, mosquitoes and biting blackflies (Dulmage, 1981). A really impressive discovery, however, would be a neurotoxin or toxin acting by external contact with the insect.

In the UK successful control of insects with fungi has been achieved in the glasshouse. Notably, *Verticillium lecanii* can be used to control aphids, scale insects, thrips and red spider mite (Hall, 1981; Gillespie *et al.*, 1982). World wide, a few fungi have achieved success in field-crop situations: species of *Beauveria* have been developed to control Colorado beetle and the codling moth, and species of *Metarhizium* control spittlebugs (Ferron, 1981). The use of fungi in control of glasshouse and agricultural pests is discussed by Quinlan and Gillespie in Chapters 2 and 3, respectively. Many questions have arisen, notably the importance of the speed of spore germination, the speed of growth and the extent of sporulation of different strains of different potency.

Like plants, insects will become diseased only in a well defined set of environmental conditions, appropriate moisture availability and temperature being the predominant factors. These conditions must, therefore, be carefully considered when employing biological control agents, in the glass-house or in the field. However, strains vary in response to these factors. Could strain selection and manipulation improve the performance in terms of practical field control?

1.2.2.3 *Nematode parasites.* Fungi that parasitise eelworms are found in all the major subdivisions of nematodes. Endoparasitic forms such as *Meria* and *Hirsutella* produce sticky spores which attach to the cuticle of the eelworm, leading to penetration and invasion (Kerry & Crump, 1980). Many others (e.g. *Dactylella oviparasitica*) are known to infect the egg stage (Stirling & Mankau, 1979). Nematodes live in the relatively clois-tered environments of the soil and within plant roots, thereby presenting a major problem in terms of the successful targeting of either predaceous or parasitic fungi for their control. Such problems are discussed by Stirling (Chapter 10).

1.2.2.4 *Parasites of fungi: mycoparasitism.* Fungi frequently grow on other fungi in nature, although association does not necessarily imply a parasitic relationship. Fungi of different species can grow in mutual as-sociation, sharing the same food base, but without one partner apparently affecting the well-being of the other. A true parasite–host relationship does exist between many fungi in which parasite relies directly on host for its maintenance. Madelin (1968) has reviewed fungi parasitic on other fungi and lichens. As with other types of parasite, there are obligate and facultative forms of association. Most are obligate: since parasitic fungus depends on host for its continued development, the effects on the host are often negligible. Consequently, they may be of little value for biological control of fungal plant pathogens and anyway they cannot be cultured on artificial media. Examples of obligate mycoparasites occur among the chy-trids, the Mucorales and the Fungi Imperfecti and being biotrophic they are specific to single host species or to a few closely related host species, which may be taxonomically similar to or very different from the parasite. For example, more than 80 fungi parasitise rusts and powdery mildews (Kranz, 1981): the biotrophic imperfect fungus *Darluca filum* parasitises over 360 host species (Eriksson, 1966).

In contrast, the necrotrophic, facultative, fungal parasites destroy their hosts by means of enzymes and toxins, and continue to absorb nutrients from the dead host cells. Some also physically impede growth and repro-duction of the host (Deacon, 1983). Their host range is relatively wide and they can be cultured readily on artificial media. These fungi include several Fungi Imperfecti and wood-rotting Basidiomycetes. They penetrate direct-ly through the host hyphal wall, either with or without previously coiling around the hyphae. Such fungi have great potential in the control of fungal plant pathogens, particularly soil-borne species. Notable examples are *Trichoderma harzianum* to suppress damping-off fungi, and *Coniothyrium minitans* which — under field conditions — successfully attacks the sclerotia of pathogens in the genus *Sclerotinia*. A discussion of the potential for mycoparasitism to suppress plant disease is provided by Whipps and Cooke (Chapter 10) in this text.

1.2.3 *Antagonistic fungi*

The ability of a fungus to exist in a particular habitat such as the soil or on the surface of a plant organ is partly determined by its ecological relationships with other microorganisms. These interrelationships are often antagonistic in nature. One or more of the organisms may be harmed or have their activities curtailed. This therefore provides another approach to the manipulation of fungi as biological control agents. The most useful fungi are species of *Trichoderma*, *Penicillium* and *Gliocladium* which antagonise plant pathogens on or near the surface of hosts such as seeds and seedlings and near wounds (Kommendahl & Windels, 1980), or on the phylloplane (Blakeman & Fokkema, 1982).

The production and release of a wide range of compounds including antibiotics is widespread among the fungi, occurring in relatively few of the lower fungi but commonly in many members of the Ascomycetes, Basidiomycetes and Fungi Imperfecti. They are toxic to a broad spectrum of microorganisms, especially fungi and bacteria. Production of toxic compounds may be important in conferring a competitive advantage to the producer, but other factors must also be considered, such as the relative abilities to colonise different microhabitats and utilise different substrates. For example, Marx (1972) demonstrated that ectomycorrhizae may act as biological deterrents to pathogenic root infections. Thus it may be possible to increase the population and activities of a useful antagonist by modification of its microhabitat and thereby improve its potency as a biological control agent. The practical potential for this in the soil and on the phylloplane is discussed in this book by Faull in Chapter 7 and by Blakeman in Chapter 8.

1.3 **Fungi as targets of other biological control agents**

Although the main consideration in this book is application of fungi as *biological control agents*, we have seen that fungi, as plant pathogens, may also be *targets* of antagonistic fungi. To complete the concept of the involvement of fungi in biological control systems it is appropriate to consider them more fully in this context as target pest organisms. There are, indeed, enemies of fungi in other taxonomic categories. Viruses or mycophagous insects, for example, must occasionally exert natural constraints on fungal development, and potential may exist for their exploitation as control agents under certain conditions. Ghabrial (1980) indicated that 'myco-viruses', i.e. viruses with potentially adverse effects on phytopathogenic fungi, might be exploited as a means of biological control. Fungal pathogens of plants known to be susceptible to disease caused by viruses include *Helminthosporium victoriae*, the cause of Victoria blight of oats (Lemke & Nash, 1980), and *Gaeumannomyces graminis*, the cause of take-all of wheat (Rawlinson *et al.*, 1973).

More evident, however, is an undoubted promise of antagonistic bacteria to control fungal pathogens on wounded plant tissue (such as leaf scars or pruning wounds) or on the phylloplane or in the rhizosphere; species of *Pseudomonas* and *Bacillus* are of particular interest. *B. thuringiensis* exhibits a double-edged offensive: apart from its well established use as an insecticide, it is also antagonistic to the bean rust fungus *Uromyces*. Examples of bacterial antagonists of fungi are included in this book (Blakemen, Chapter 9). Siderophores are compounds produced by root-inhabiting fluorescent pseudomonads and other microorganisms under low-iron stress. These iron transport molecules not only enhance plant growth, but also cause reductions in root-zone populations of fungi and bacteria, which may well include pathogens. Fungal pathogens known to be suppressed by siderophores include *Gaeumannomyces graminis* (take-all of wheat) and *Fusarium oxysporum* f. sp. *lini* (flax wilt) (Kloepper *et al.*, 1980; Leong, 1986).

An intriguing observation of the indirect influence of an insect on the reduction of the population of a fungal pathogen was made by McGregor & Moxon (1985). Ants of the species *Anoplolepis longipes* exclude other ants which harbour — in the matrix of their tents — the pathogen *Phytophthora palmivora*, causal agent of pod rot in cocoa. Introduction of this species of ant into cocoa plantations may therefore provide a unique biological control method.

1.4 Enhancement of biological control potential of fungi

A common *cri de coeur* of researchers in biological control is that, although the degree of success under controlled environmental conditions arouses optimism, trials in the field, or even in relatively predictable glasshouse conditions, may rapidly turn euphoria into despondency. Often this is true even though the utmost care may have been taken in applying the concentration of fungal propagules to the target pest, optimal for a given situation, or to its habitat, at the optimum time and under the best available climatic conditions. Depending on the situation, there may be many reasons for this. For example, candidate fungi may fail to survive the relatively severe competition encountered in field conditions; they may lack sufficient virulence to overcome weeds, insects or fungi that are not raised under experimental conditions; they may not generate or maintain sufficient inoculum at the site of attack, or simply not thrive in the rapidly fluctuating environmental conditions.

There are four ways in which these shortfalls to field success may possibly be overcome:

(a) application of chemical adjuvants to protect and/or stimulate the development of the fungus agent;
(b) application of chemical adjuvants which interfere with the defences of the pest;
(c) genetic manipulation of the fungus agent to improve its virulence to the target pest or its general vigour in the environment;
(d) manipulation of the environment, e.g. by altering the spacing of crop plants.

In attempting to develop fungal pathogens as mycoherbicides against bracken, it has been practicable to adopt the first three approaches. Propagules were mixed with alginates to improve their adhesion to the aerial surfaces of the weed and to protect them from desiccation during the critical period between germination and penetration, and with dilute nutrient solutions to stimulate their germination and growth. Addition of very dilute solutions of contact herbicides such as ioxynil (4-hydroxy-3, 5-diiodobenzonitrile) and of non-ionic surfactants such as 'Silwet L-77' helps to overcome the weed's resistance to penetration. Such additives significantly increased the number of penetration events in the field (Burge *et al.*, 1986), but invasion of bracken tissues was limited owing to the rapid production of lignified papillae as mentioned earlier. Attempts to reduce the potential for lignification in the host by the application of inhibitors of lignin biosynthesis (e.g. amino oxyacetic acid) succeeded in the laboratory but not in the field (McElwee, 1987). Significant increases in virulence of fungi to bracken in growth-cabinet conditions have been achieved by selecting fungal mutants obtained using N-methyl-N[1]-nitro-N-nitroso-guanidine (NTG) (unpublished data), but these are yet to be tested in the field. The potential for genetic manipulation of fungi to improve their performance as biological control agents is discussed by Heale (Chapter 11).

An additional phenomenon which may have an important bearing on field success is variation in susceptibility of the pest. Although it has been stated earlier that there is a relatively low risk of development of resistance to biological control agents, it is none the less probable that within pest populations there is a range of susceptibility to attack. Again this can be exemplified by our own work with bracken. This weed reproduces largely by vegetative means, ostensibly to form genetically uniform stands. Thus, a single plant may cover several hectares by virtue of the underground spread of a branching network of rhizomes, all originated from the initial plant. Although there is as yet no positive experimental evidence to indicate genetic differences between stands from two geographically different areas, or indeed within stands that have coalesced, McElwee (1987) has demonstrated that bracken from two different localities in Scotland differs greatly in susceptibility to curl-tip disease caused by *Ascochyta pteridis* and *Phoma aquilina*. This is reflected directly by the differences in induction of the resistance response, i.e. the production of lignified papillae, in the bracken from each area. Clearly, then, the susceptibility of pests from a variety of sources should be examined under laboratory conditions before carrying out extensive field trials.

1.5 Production and commercialisation of fungi as biological control agents

Lisansky (1985) has outlined the steps leading to commercialisation of pathogens. The product has to be developed in such a way as to overcome some of the constraints limiting success in the field. Formulation of fungal pathogens includes methods to stabilise them for as long as practicable and also to make them as immune as possible from 'user abuse'. This is necessary since biological control agents are much more sensitive to conditions of storage and application than chemical pesticides and are therefore vulnerable to criticism when performance is poor as a result of mishandling. Many questions await research. It is obvious also that virulence of fungi, like that of any biological control pathogen, must not be lost during its maintenance on culture media, during large-scale production procedures, or during storage. Repeated transfer of entomogenous fungi, for example, can result in loss of virulence and sometimes even pathogenicity (Muller-Kogler, 1965). Virulence can be restored only by repeatedly passing the fungus through the host. Commercially, such problems can be minimised by storing a large number of aliquots of a single spore isolate deep-frozen or freeze-dried.

Commercially, fungal propagules are produced by fermentation, either in 'submerged' (fully liquid) systems or 'semi-solid' ones where the fungus is raised on the wet surface of a solid material. The method used depends on the fungus, but where possible the submerged system is used since it is easier to regulate the growing conditions.

Production of propagules is followed by harvesting, i.e. 'down-stream processing'. This is often hazardous to the organism because of the sudden change in conditions that can occur. For example, it is necessary to take account of potential osmotic shock, extreme temperatures, or desiccation. After harvesting, the material is formulated with suitable adjuvants with a variety of functions such as preservation of fungal propagules, allowing them to be tank-mixed and sprayed on to the host or into its habitat and to be retained there in an active state for as long as possible. Before marketing, however, the specification must accurately state the nature and minimal purity of the material. An essential feature is safety testing, an aspect well monitored by national governments. These include tests for potential toxicity or illicitation of allergic responses. A strategy for evaluating the safety of organisms for biological weed control has been put forward by Wapshere (1974). After safety clearance, the biocontrol organism can be registered in all the countries of its intended use (a procedure which may take about two years) and batches can be marketed following suitable quality control testing. The production and formulation of fungi — a challenging subject — is discussed for mycoinsecticides by Bartlett & Jaronski (Chapter 4).

1.6 Fungi in integrated control programmes

Integrated pest management (IPM) involves, simultaneously or sequentially, the use of several methods of control (Huffaker, 1980). This approach, as well as being economical, ensures the maintenance of environmental quality because it allows significant reductions to be made in the application of pesticides. Biological control agents are particularly valuable components of IPM systems. Throughout the world, programmes are being developed for human health (mosquitoes), numerous crops (alfalfa, cotton, apples, citrus fruits, peaches, pears, grapes, oil palms, rubber and cocoa), and animal husbandry (sheep, cattle and poultry) (Huffaker, 1971, 1980). In each of these programmes, preservation and/or augmentation of natural enemies has been a critical factor. Klassen (1981) has pointed out that documented evidence exists on the management of plant pathogens, plant parasitic nematodes, arthropods and weeds by joint use of biological control agents together with cultural practices, pesticides or resistant cultivars. Often, the potential for integrating biological control and other control methods is discovered by serendipity: Chinn (1971) found that spores of the fungal pathogen *Helminthosporium sativum* were not affected by 5 ppm methyl mercury dicyandiamide, but the treatment stimulated an increase in the soil of certain microorganisms (especially *Penicillium* spp.) which were thought to be responsible for the disease control obtained. Such observations indicate the potential for selective inhibition or stimulation of organisms or groups of organisms by the judicious application of chemicals. The role of fungi in integrated control programmes for the control of plant diseases is reviewed by Papavizas & Lewis in the last chapter of this text.

A well known adverse effect of pesticides is the destruction of beneficial organisms as well as the target pest. This increases the need for pesticidal application to destroy those pests that would have otherwise been held in check by the controlling organisms. Of course, when pesticides are purposefully used in conjunction with biological control agents, it is necessary to appreciate fully the effect of the pesticide on the biological agents. Such evidence has been compiled for several biological control agents: for example, Hassan & Oomen (1985) and Ledieu (1985) have assembled tables indicating the compatibility of a large number of pesticide formulations with six different beneficial organisms employed in the greenhouse. The organisms include the fungus *Verticillium lecanii*, the agent in mycoinsecticide used against aphids and scales. It would appear from these tables that the fungus may be used satisfactorily with all the insecticides and acaricides tested. Fungicides such as benomyl, captan, chlorothalonil, dichlofluanid, imazalil and maneb are incompatible when used near to the time of spraying with *V. lecanii* and should be applied seven days after or (preferably) seven days before application of *V. lecanii*. Several others such as dinocap, fenarimol, iprodione, oxycarboxin, pyrazophos, triforine and vinclozolin were found to be compatible except when applied together with *V. lecanii* in the same spray.

1.7 Research effort in the UK

In order to obtain some concept of the amount of research devoted to biological control in the UK a survey of projects underway in 1985–1986 in British Universities, polytechnics and colleges and in government and other institutions has been carried out by this author. This was completed by examination of published works of reference and by correspondence. In all, 52 projects were detected and in 50 of these the general or specific nature of the controlling agent and of the target pest were specified in the project titles (Table 1.1). It is unlikely that the survey is complete, but it does indicate clearly the importance and value of fungi as potential biological control agents because they are involved as such in 28 (56%) of the projects indicated. Most of the research (69%) was taking place at universities, with about 15% each at polytechnical colleges and at research institutes. Approximately half of the projects attracted government finance but industrial participation was only apparent in five (1%).

The paucity of industrial involvement may be indicative of two things: (1) the small but growing number of actively participating companies, and (2) the probability that many of the projects are, for one or more of the reasons stated above, a long way from commercialisation. This is perhaps understandable in the present economic climate, and companies are bound to favour research projects which promise expeditious returns. However, the pressures of certain environmentally conscientious governments and organisations are steadily increasing, and given the benefit of financial

Table 1.1. Numbers of UK biological control projects concerned with different agents and target pests (1985–86)

Control agents	Target pests				Total
	Fungi	Insects and mites	Eelworms	Weeds	
Fungi	12	11	3	2	28
Insects and mites	0	5	1	1	7
Eelworms	0	3	0	0	3
Bacteria	2	4	0	0	6
Viruses	1	5	0	0	6
Total	15	28	4	3	50

encouragement many of the hard won findings outlined in this book will be brought to fruition.

References

Albersheim, P. & Anderson-Prouty, A.J. (1975). Carbohydrates, proteins, cell surfaces and the biochemistry of pathogenesis. *Annual Review of Plant Physiology*, **26**, 32–52.
Blakeman, J.P. & Fokkema, N.J. (1982). Potential for biological control of plant diseases on the phylloplane. *Annual Review of Phytopathology*, **20**, 167–92.
Boman, H.G. (1981). Insect responses to microbial infections. In: *Microbial Control of Pests and Diseases, 1970–1980*, ed. H.D. Burges, pp. 769–84. Academic Press, London.
Burge, M.N. & Irvine, J.A. (1985). Recent studies on the potential for biological control of bracken using fungi. *Proceedings of the Royal Society of Edinburgh*, **86b**, 187–94.
Burge M.N., Irvine, J.A. & McElwee, M. (1986). The potential for biological control of bracken with the causal agents of curl-tip disease. In: *Bracken: Ecology, Land Use and Control Technology*, eds R.T. Smith & J.A. Taylor, pp. 453–7. Parthenon Publishing, Carnforth, Lancashire.
Burges, H.D. ed. (1981). *Microbial Control of Pests and Diseases 1970–1980*. Academic Press, London.
Burges, H.D. & Hussey, N.W. eds (1971). *Microbial Control of Insects and Mites*. Academic Press, London.
Charudattan, R. & Walker, H.L. eds (1982). *Biological Control of Weeds with Plant Pathogens*. Wiley, Chichester.
Chinn, S.H.F. (1971). Biological effects of Panogen PX in soil on common root rot and growth responses of wheat seedlings. *Phytopathology*, **61**, 98–101.
Cook, R.J. & Baker, K.F. (1983). *The Nature and Practice of Biological Control of Plant Pathogens*. American Phytopathological Society, St. Paul, Minn.
Deacon, J.W. (1983). *Microbial Control of Pests and Diseases*. Van Nostrand, New York.

Duddington, C.L. (1957). *The Friendly Fungi.* Faber & Faber, London.
Duddington, C.L. (1973). Zoopagales. In: *The Fungi, Vol. IVB,* pp. 231–4. Academic Press, New York.
Dulmage, H.T. (1981). Insecticidal activity of isolates of *Bacillus thuringiensis* and their potential for pest control. In: *Biological Control of Pests and Plant Diseases,* ed. H.D. Burges, pp. 193–222. Academic Press, London.
Eriksson, O. (1966). On *Endarluca caricis* Fr. O. Eriks., comb. nov., a cosmopolitan uredinicolous Pyrenomycete. *Botanica Notis,* **119,** 33–69.
Fast, P.G. (1981). The crystal toxin of *Bacillus thuringiensis.* In: *Microbial Control of Pests and Plant Diseases 1970–1980,* ed. H.D. Burges, pp. 223–48. Academic Press, London.
Ferron, P. (1981). Pest control by the fungi *Beauveria* and *Metarhizium.* In: *Biological control of Pests and Plant Diseases,* ed. H.D. Burges, pp. 465–82. Academic Press, London.
Fletcher, W.W. (1974). *The Pest War.* Blackwell, Oxford.
Fox Wilson, G. (1960). *Horticultural Pests — Detection and Control,* 2nd ed. Crosby Lockwood, London.
Ghabrial, S.A. (1980). Effects of fungal viruses on their hosts. *Annual Review of Phytopathology,* **18,** 441–61.
Gillespie, A.T., Hall, R.A. & Burges, H.D. (1982). Control of onion thrips, *Thrips tabaci* and the red spider mite *Tetranychus urticae* by *Verticillium lecanii.* Proceedings of the IIIrd International Colloquium of Invertebrate Pathology, Papers, 101.
Gram, E.P., Bovien, P. & Stapel, C. (1969). *Recognition of Diseases and Pests of Farm Crops,* 2nd edn, Blandford Press, London.
Hall, R.A. (1981). The fungus *Verticillium lecanii* as a microbial insecticide against aphids and scales. In: *Microbial Control of Pests and Plant Diseases 1970–1980,* ed. H.D. Burges, pp. 483–98. Academic press, London.
Hassan, S.A. & Oomen, P.A. (1985). Testing the side effects of pesticides on beneficial organisms by OILB working party. In: *Biological Pest Control,* eds N.W. Hussey & N. Scopes, pp. 145–52. Blandford Press, Poole, Dorset.
Higgins, M.L. & Pramer, D. (1967). Fungal morphogenesis: ring formation and closure by *Arthrobotrys dactyloides. Science,* **55,** 345–6.
Huffaker, C.B. ed. (1971). *Biological Control.* Plenum Press, New York.
Huffaker, C.B. ed. (1980). *New Technology of Pest Control.* Wiley Interscience, New York.
Iffland, D.W. & Allison, P.V.B. (1964). Nematode trapping fungi: evaluation of axenic healthy and galled roots as trap inducers. *Science,* **146,** 547–8.
Irvine, J.I.M., McElwee, M. & Burge, M.N. (1987). Association of *Phoma aquilina* and *Ascochyta pteridis* with curl-tip disease of bracken. *Annals of Applied Biology,* **110,** 25–31.
Jones, F.G.W. & Jones, M. (1974). *Pests of Field Crops,* 2nd edn. Edward Arnold, London.
Kerry, B.R. & Crump, D.H. (1980). Two fungi parasitic on females of cyst nematodes (*Heterodera* spp.). *Transactions of the British Mycological Society,* **74,** 19–25.
Klassen, W. (1981). The role of biological control in integrated pest management systems. In: *Biological Control in Crop production,* ed. G.C. Papavizas, pp. 433–45. Allanheld & Osmun, Totowa, N.J.
Kloepper, J.W., Leong, J., Teintze, J. & Schroth, M.N. (1980). *Pseudomonas*

siderophores: a mechanism explaining disease suppressive soils. *Current Microbiology*, **41**, 317–20.

Kommendahl, T. & Windels, C.E. (1980). Introduction of microbial antagonists to specific courts of infection: seeds, seedlings and wounds. In: *Biological Control in Crop Production*, ed. G.C. Papavizas, pp. 227–48. Allanheld & Osmun, Totowa, NJ.

Kranz, J. (1981). Hyperparasitism of biotrophic fungi. In: *Microbial Ecology of the Phylloplane*, ed. J.P. Blakeman, pp. 327–52. Academic Press, London.

Kranz, J., Schmutterer, H. & Koch, W. (1979). *Diseases of Pests and Weeds in Tropical Crops*. Wiley, Chichester.

Kurstak, E. ed. (1982). *Microbial and Viral Pesticides*. Marcel Dekker, New York.

Lawton, J.H. (1986). Biological control of bracken. In: *Bracken: Ecology, Land Use and Control Technology*, eds R.T. Smith and J.A. Taylor, pp. 445–52. Parthenon Press, Carnforth, Lancashire.

Ledieu, M. (1985). Evaluation of side effects of pesticides by the Glasshouse Crops Research Institute. In: *Biological Pest Control*, eds N.W. Hussey & N. Scopes, pp. 153–61. Blandford Press, Poole, Dorset.

Lemke, P.A. & Nash, C.H. (1980). Fungal viruses. *Bacteriological Review*, **38**, 29–56.

Leong, J. (1986). Siderophores: their biochemistry and possible role in the biological control of plant pathogens. *Annual Review of Plant Pathology*, **24**, 187–209.

Lisansky, S.G. (1985). Production and commercialisation of pathogens. In: *Biological Pest Control*, eds N.W. Hussey & N. Scopes, pp. 210–18. Blandford Press, Poole, Dorset.

Luttrell, E.S. (1974). Parasitism of fungi on vascular plants. *Mycologia*, **66**, 1–15.

Madelin, M.F. (1968). Fungi parasitic on fungi and lichens. In: *The Fungi, Vol. III*, eds G.C. Ainsworth & A.S. Sussman, pp. 253–69. Academic Press, London.

Marx, D.H. (1972). Ectomycorrhizae as biological deterrents to pathogenic root infections. *Annual Review of Phytopathology*, **10**, 429–54.

McElwee, M. (1987). Studies on the Potential Development of the Fungal Pathogens *Ascochyta pteridis* and *Phoma aquilina* as Agents of Biological Control of Bracken. PhD thesis, Strathclyde University, Glasgow.

McGregor, A.J. & Moxon, J.E. (1985). Potential for biological control of tent building species of ants associated with *Phytophthora palmivora* pod rot of cocoa in Papua New Guinea. *Annals of Applied Biology*, **107**, 271–7.

Muller-Kogler, E. (1965). *Pilzkrankheiten bei Insekten*. Paul Parey, Berlin.

Rawlinson, C.J., Hornby, D., Pearson, V. & Carpenter, J.M. (1973). Virus-like particles in take-all fungus *Gaeumannomyces graminis*. *Annals of Applied Biology*, **74**, 197–209.

Stapely, J.H. & Gayner, F.H.C. (1969). *World Crop Protection, Vol. 1. Pests and Diseases*. Iliffe Books, London.

Stirling, G.R. & Mankau, R. (1979). Mode of parasitism of *Meloidogyne* and other nematode eggs by *Dactylella oviparasitica*. *Journal of Nematology*, **11**, 282–8.

Taylor, J.A. (1986), The bracken problem: a local hazard and a global issue. In: *Bracken: Ecology, Land Use and Control Technology*, eds R.T. Smith & J.A. Taylor, pp. 21–42. Parthenon Press, Carnforth, Lancashire.

Templeton, G.E. & Greaves, M.P. (1984). Biological control of weeds with fungal pathogens. *Tropical Pest Management*, **30**, 333–8.

Walker, H.L. & Templeton, G.E. (1978). *In vitro* production of a phytotoxic metabolite by *Colletotrichum gloeosporioides* f. sp. *aeschynomene*. Plant Science Letters, **13**, 91–6.

Wapshere, A.J. (1974). A strategy for evaluating the safety of organisms for biological weed control. *Annals of Applied Biology*, **77**, 201–11.

Windels C.E. & Lindow, S.E. (1985). *Biological Control on the Phylloplane*, Symposium Book No. 3. American Phytopathological Society, St. Paul, Minn.

Use of fungi to control insects in glasshouses

2.1 Introduction

2.1.1 *The stimulus*

The glasshouse provides an almost ideal environment for insect growth and reproduction. There is usually ample warmth, food and water. The crop plants have been bred for ultimate yield, which as a side effect usually involves minimal protection against insect species. In the absence of any action, the glasshouse grower will always produce a crop infested with a variety of insect and mite pests and these will directly or indirectly, sooner or later, cause yield reduction or total crop loss.

The need for good, predictable protection against pests has meant heavy reliance on pesticides. Unfortunately, because of the rapid multiplication rate of these pests in glasshouse conditions, 'superbugs' resistant to an array of pesticidal compounds have developed, leading to an inability to control many pests. This has been particularly so with the red spider (two-spotted) mite *Tetranychus urticae*, with the glasshouse whitefly *Trialeurodes vaporariorum* in the European tomato and cucumber industry, and to a lesser extent with aphids on a variety of crops.

The rate of development of resistance to chemicals by some of these insects can be dramatic. In the year that pyrethroid insecticides were introduced into the United Kingdom tomato industry for control of the glasshouse whitefly, many growers reported a need to apply higher rates by the end of a 4-month season. Recent reports indicate that the glasshouse whitefly in many areas may require up to 2000 times more pyrethroid to achieve results equivalent to those obtained when the insecticide was first introduced 3 years earlier.

In addition, heavy insecticide use has produced problems of phytotoxicity and of pesticide residues. Tomato growers have reported up to 20% increase in yield by changing from a chemical insecticide programme to an integrated programme using predators and parasites. This phytotoxicity, often overlooked by growers, is due to a variety of effects including induced flower drop. Additionally, pesticide residues from the more persistent compounds or those with systemic activity often limit the use of many pesticides in continuously harvested crops such as cucumbers or tomatoes.

There has therefore been an increasing interest by growers in using biological control agents within integrated pest management programmes. Approximately 80% of the UK cucumber industry and 50% of the tomato industry now use biological control agents for spider mite and whitefly control. However, in other crops, the changeover has been slower; only 5–10% of the UK chrysanthemum industry uses a programme developed by the UK Agricultural Development and Advisory Service (ADAS). In continental Europe, for example in the recently developed protected cultivation industry in Almeria, Spain, the use of biological control is negligible.

The slow or incomplete change to integrated programmes highlights the often inadequate or unpredictable control provided by many biological agents. Thus the development of new control measures is still required on some crops, and alternatives to chemicals are being desperately sought, for example with glasshouse whitefly on tomatoes or leaf miner control on both chrysanthemums and tomatoes.

The search for new biological control agents has covered the spectrum of parasites, predators, pathogenic nematodes, bacteria, fungi, viruses and behaviour-modifying compounds such as pheromones. Notable in this list are entomopathogenic fungi, which have been and still are one of the top priorities in the search for new control agents.

2.1.2 *Use of entomopathogenic fungi*

The glasshouse industry has often been considered highly suitable for the development and application of entomopathogenic fungi. This view has been held because glasshouse crops are among the most intensively and technically cultivated and the most expensive and pest-susceptible in the world. In other words, glasshouse growers are technically minded, innovative, and familiar with serious, predictable pest problems involving regular prophylactic pest control. Additionally a high crop value can support the relatively high cost of a new, small-market and highly specific pest control agent.

The factor that has made the glasshouse appealing to the advocate of fungal insect control has been the control of temperature and humidity conditions within the crop. Glasshouse crops commonly have a higher relative humidity and more moderate night-time temperature than those in field crops from the same region. Additionally, the modern grower can monitor and adjust the environmental conditions within the crop using computers which can increase humidity, alter temperature, improve or restrict ventilation, and alter supplies of nutrients and water to plants at precise times on given occasions as required. Thus, fungi are not restricted by variability of the weather.

Although the glasshouse environment is maintained primarily for the optimal growth of plants, minor changes can be made to suit insect-pathogenic fungi. For example, the grower may be able to water the crop using overhead spray lines simulating rain and providing ample free water for fungal growth. In chrysanthemums, normal practice particularly favours the fungus because the plants are commonly covered with a low-level black screen for 10–12 hours a day in order to control flowering time; or with 'thermal screens' designed to save heat. Both practices increase humidity.

The requirement of fungal pathogens for periods of moderate tempera-ture and high humidity for spore germination, colonisation, killing and growing on their hosts is well documented (e.g. Hall, 1981; Wilding, 1981). Although such conditions can be found in field crops, the unpredictable and generally drier and cooler conditions (in northern Europe at least) militate against successful use of fungi as microbial pesticides. For this reason, there are currently no commercial fungal pesticides available for insect control on outdoor crops in northern Europe or North America. 'Mycar', a formulation of the fungus *Hirsutella thompsonii* (produced by Abbott Laboratories), was available commercially in 1981 for control of the citrus rust mite, *Phyllocoptruta oleivora* in Florida citrus groves. However, in its first summer of commercial use, the weather was 'hot and dry' — not conducive to good *Hirsutella* growth — and disappointing results were obtained (C.W. McCoy, personal communication). Mycar has not been widely available since.

2.1.3 *Constraints*

Despite the high pressure on some growers to move towards integrated control and hence possibly to fungal agents, worldwide the move has been slow. The key reason is that such pressure has not yet reached most growers outside the UK. Chemicals remain mostly effective and cheap. Until this situation changes, or until some other constraint appears (e.g. legislation or a market demand for chemical-free produce), then integrated pest management will remain regional and uncommon. Additionally, fungi are expensive to produce commercially on a small scale and hence expensive to buy and use. They have a very short shelf-life, needing cold storage. They are usually very specific in activity, which means (1) that there is a more limited market and hence less commercial interest in them, and (2) that the grower has to use other control agents for other pests. Fungi need to be used precisely in terms of application and integration; the periods of high humidity that they require conflict with a grower's inclination to reduce humidity to order to restrict growth and infestation by plant pathogens. Fungi need careful post-application treatment in maintaining the humidity and avoiding deleterious fungicides. The fungi often take a long time to kill their host, up to 2–3 weeks compared with 2–3 hours for some chemical pesticides.

These technical constraints necessitate that the grower should have a greater understanding of the pest problems and the control agents being used. They also require some patience and much belief, particularly with prophylactic use or where living insects can be seen on the crop for a week or two after application of the fungus. There is also a need for greater commitment, understanding and profit margin for the salesmen and marketing outlets, because more sales support in terms of explanations and follow-up visits is necessary as compared with a synthetic chemical. All these factors work against the use of fungal insecticides in the commercial glasshouse, whatever the technical merits of the fungus in the hands of the scientist.

2.2 **Pests and pathogens**

A list of arthropod pests against which control measures may have to be taken in the UK glasshouses is shown in Table 2.1. There are similar lists for the rest of northern Europe and for most of the world. The key insect pests are: aphids, whiteflies, spider mites, caterpillars and leafminers. Although others, such as thrips, vine weevil, mealybugs, leafhoppers, etc., can (and have) become important locally on particular crops, the key insects are normally found on a range of plants and have been at the centre of development of biological pest control methods.

The development of insect-pathogenic fungi in glasshouses can be seen

Table 2.1. Pests of glasshouse crops in the United Kingdom (from Morgan & Ledieu, 1979)

Pest	Species example	Crops
Aphids	*Myzus persicae, Aphis gossypii Nasanovia ribisnigri, Aulacorthum*	O, B, Cu, L, P, T
Capsids	*Lygus rugulipennis*	O
Caterpillars	*Cacoecimorpha, Phlogophora, Lacanobia*	O, T
Earwigs	*Forficula auricularia*	O
Flies	*Psilla rosae, Philophylla, Sciara*	C, Cu
Leafhoppers	*Zygina pallidifrons*	O, Cu, T
Leafminers	*Phytomyza, Liriomyza trifolii*	O, P, T
Mealybugs	*Planococcus citri, P. adonidum*	O, G
Millipedes	*Oxidus gracilis*	Cu
Mites	*Tetranychus urticae, T. cinnabarinus*	O, B, Cu, G, P, T
Scale	*Parthenolecanium, Coccus, Aspidiotus*	G, O
Slugs		O, C, L, T
Springtails	*Collembola* sp.	Cu, L, T
Symphylids	*Scutigerella immaculata*	O, B, Cu, L, T
Thrips	*Heliothrips femoralis, Thrips tabaci*	O, Cu, P, T
Vine weevils	*Otiorhynchus sulcatus*	O
Whitefly	*Trialeurodes vaporariorum*	O, B, Cu, P, T
Wireworms	*Agriotes obscurus*	O, B, L, T
Woodlice	*Armadillidium, Oniscus, Porcellio*	O, Cu, P, T

Key: B, beans; C, celery; Cu, cucurbits (e.g. cucumber); G, grapevine; L, lettuce; O, ornamentals; P, sweet pepper and aubergine; T, tomato.

as an interaction between various forces, namely failure of chemical pesticides, pathogen availability, technical problems, success of alternative biological agents, and market need. Thus, successful development of the mite predator, *Phytoseiulus persimilis*, has limited the development of pathogens for the control of the two-spotted or red spider mite, *Tetranychus urticae*. Despite this, promising early results in the laboratory and in the glasshouse have been obtained with *Verticillium lecanii* (Gillespie *et al.*, 1982a), *Neozygites floridana* (Brandenburg & Kennedy, 1983) and *Hirsutella thompsonii* (Gardner *et al.*, 1982). Similarly, the bacterial pathogen of lepidopteran caterpillars, *Bacillus thuringiensis* (Bt) — commercially available in a number of formulations, has restricted development of alternative caterpillar pathogens. Bt is cheap, safe, effective and more resilient than most fungal agents. Its replacement in a biological programme would be difficult. However, some fungi, such as *Nomuraea rileyi*, have been investigated because of their ability to spread within a population and hence give prolonged control (Ignoffo, 1981).

The biology of an insect may make the use of entomopathogens extremely difficult. For example, the North American leafminer, *Liriomyza trifolii*, with its larval stages inside the leaf, is protected from fungal infection. For this reason, development has been on hymenopterous parasites such as *Opius* and *Dacnusa*.

Most research and development, and the only operational commercial production and marketing effort, has been on the homopterous insects: aphids (e.g. *Myzus persicae* and *Aphis gossypii*) and whiteflies (*Trialeurodes vaporariorum*). These insects are both numerous on a wide range of crops, and have developed resistance to chemical pesticides extremely rapidly. Extensive programmes have investigated the potential for the use of parasites and predators, and this group of pests has been the subject of by far the most activity in terms of developing fungal pathogens.

The glasshouse whitefly, *Trialeurodes vaporariorum*, is a major pest on most glasshouse crops, especially tomatoes and cucumbers. Large populations of several hundred adults per plant can build up quickly and have to be treated. The whitefly causes damage primarily by deposition of honeydew — the sugary secretion of its nymphs — on to the leaves and fruit which then supports growth of black-spored fungi, termed sooty mould. In large numbers, however, whiteflies will damage the plant directly.

The whitefly larval parasite, *Encarsia formosa*, is now widely available throughout the world and is used extensively in the glasshouse industry. Although it is generally successful, failure of the parasite in some circumstances has resulted in continued investigation into pathogenic fungi. Various species have been investigated, notably *Verticillium lecanii* (e.g. Hall, 1982a; Quinlan, 1982; Kitizawa *et al.*,1984); *Aschersonia aleyrodis* (e.g. Ramakers *et al.*, 1982; Adam, 1983; Landa, 1984) and *Beauveria bassiana* (Borisov & Vinokurova, 1983; Treifi, 1984). The most successful of these fungi has been *Verticillium lecanii*, which was developed commercially by Microbial Resources Ltd (MRL) with the Glasshouse Crops Research Institute into a microbial insecticide, 'Mycotal' (Rights to all MRL's production and formulation technology have been bought by Novo Industri, Denmark.) In comparison with the other candidates, *V. lecanii* is more effective, cheaper to produce, infective to adult as well as larval whiteflies and spreads well within the pest population. *Aschersonia*, although as infective and probably faster in action, is more expensive to produce, non-infective to adults and does not induce epidemics or 'epizootics' in the way that *Verticillium* does. Ramakers *et al.* (1982), for example, needed to apply a suspension of 2.8×10^{12} conidia per hectare at weekly intervals in order to obtain 71% infection of whitefly nymphs on cucumbers. In contrast, Quinlan (1983) obtained 99–100% infection of both nymphs and adults following a single application of 5×10^{11} spores per hectare of *V. lecanii* to a commercially grown tomato crop. Similar results were found by Hall (1982b). *V. lecanii* is discussed in more detail below.

Aphids are less of a problem in the glasshouse than whiteflies, primarily because they are not regular serious pests on the two main glasshouse vegetable crops, tomatoes and cucumbers. In the UK, aphids are primarily a pest on ornamental crops, most notably chrysanthemums. Damage is caused by honeydew deposition, distortion of plant growth and general cosmetic damage to cut flowers or pot plants.

Various biological means of controlling aphids have been investigated. Parasites and predators, notably *Aphidius matricariae* (a parasite of *Myzus persicae*) and *Aphidoletes aphidomyza* (a predator), have been extensively studied and can be purchased from commercial biological control suppliers. These have been used with only limited success, the specificity of *A. matricariae*, for example, making little impact on aphid populations of mixed species. Aphid populations, ranging widely in species, are, however, commonly found infected with a variety of fungal species, and this has led to the development of *Verticillium lecanii* as the microbial insecticide Vertalec, as well as to continued investigation into a variety of other organisms. The use of Vertalec is discussed below. In addition other fungi, such as *Erynia neoaphidis* (Wilding, 1983), are being investigated not only in the glasshouse but also for field use.

The commercial pressure, or market need, to develop other pathogens has been less acute and little or no activity has been forthcoming. The potential, however, is evident in a number of cases.

Thrips (e.g. *Thrips tabaci*) are known to be infected by a variety of fungal species, including *Verticillium lecanii* (Hall, 1982b; Gillespie *et al.*, 1982a), and populations in sweet pepper crops under glass in Holland have been infected with *Entomophthora thripidum* (Samson *et al.*, 1979). Various authors (e.g. Keller & Wuest, 1983) have also reported that *Neozygites parvispora* infects *Thrips tabaci* on, for example, onions in Switzerland. None of these pathogens has been introduced commercially.

Leafhoppers are susceptible to a wide range of fungi. Gillespie *et al.* (1982b) demonstrated susceptibility to *Metarhizium anisopliae* and *Verticillium lecanii*. Marletto & Maggiora (1983) isolated no fewer than six species from *Zyginidia pullula*, finding *Cladosporium cladosporioides* to be the most infective.

Black vine weevil, *Otiorhynchus sulcatus*, a major pest on strawberries and potted plants under glass, is highly susceptible to *Metarhizium anisopliae* (Soares *et al.*, 1983; Zimmerman, 1984). Soares *et al.* (1983) obtained 95% control of weevils following an application of *Metarhizium* at 4×10^4 conidia per square centimetre of soil surface. The commercial potential of the fungus for this use should be fairly high, although the use of entomopathogenic nematodes for vine weevil control is already commercial in Holland.

The likelihood of development and eventual commercialisation of these and of any other fungi will depend upon market demand. In general, the pesticide market within the glasshouse industry is too small to support the financial recoupment required for the development of a chemical pesticide specifically for that use. Therefore, product development for these markets must either be associated with a product for a larger market, such as a field crop pest, or with an active ingredient which is relatively inexpensive for manufacturers to develop. The approach taken by registration authorities towards biological agents is helpful. Notably in the United States and in the UK no registration is required for parasites and predators, and a stepwise toxicology approach towards microbial pesticides means that non-toxic materials are exempted from long-term sub-acute toxicology studies. As a result the glasshouse industry can be a useful market niche for biological agents. The development costs are such that with biological agents comparatively smaller markets can be considered than those with new synthetic chemicals. This is particularly so if government research bodies and advisory services maintain their present involvement in the study of these organisms.

2.3 Mycotal and Vertalec

2.3.1 *Verticillium lecanii*
Mycotal and Vertalec are commercial products developed by Microbial Resources Ltd and based on different isolates of *Verticillium lecanii*. They are now made and marketed by Novo Industri, Demark, and not only represent the sole fungal insect pathogens sold for commercial use in the glasshouse industry, but also are the only products of this type on sale in the Western world for any insecticidal function. Vertalec was introduced for aphid control on chrysanthemums in 1981, and Mycotal for glasshouse whitefly control on cucumbers and tomatoes in 1982.

V. lecanii is a commonly occurring Deuteromycete fungus, well documented as a widespread saprophyte and pathogen. It has been isolated from every continent in the world and has been found growing in soils, on rotten wood, on baker's yeast, and maize seeds, and parasitising rusts as well as attacking numerous arthropods — including Hemiptera, Lepidoptera, Diptera, Hymenoptera, Collembola and eriophyid mites.

V. lecanii is non-fastidious in its dietary requirements and grows on a wide variety of media, e.g. potato dextrose, malt extract, and Sabouraud's and Czapek-Dox agars (Hall, 1981). It also grows on a variety of grains and in conventional submerged media (Hall & Burges, 1979). Its taxonomy has been described by Samson (1981) and its general biology by Hall (1981).

V. lecanii infects susceptible species of insects following the adhesion of viable spores on the cuticle. The spores germinate and grow, producing hyphae that penetrate into the body cavity where they infect and destroy the tissues. *V. lecanii* probably does not produce a toxin in the fashion of *Metarhizium anisopliae* (Roberts, 1981); its toxicity is probably based solely on proliferation within insect tissues. Even before the insect dies, the fungus grows through the insect cuticle and on the outside of the body produces sticky heads of spores which spread the infection to other insects, creating an epizootic (Fig. 2.1). At high humidities, infected insects become enveloped in white fluffy hyphae and are easily detected in the field (Fig. 2.2).

Commercially *V. lecanii* is produced by submerged fermentation. The spores are harvested and formulated to overcome problems of spore longevity. At 4°C, unformulated spores have a half-life of less than a week compared with approximately a year for formulated spores. Both Mycotal and Vertalec are supplied as dry wettable powder formulations with a count of viable colony-forming units (which usually equates to viable spores) in excess of 2×10^8 per gram of product. These products have been subjected to the full range of toxicology tests required for registration and approval by various authorities around the world. In addition, quality control procedures on batches of product released for sale include extensive microbiological purity checks, as well as subcutaneous injection of 5×10^6 viable *V. lecanii* spores into each of five mice.

2.3.2 *General principles for use*

Vertalec contains the spores of a specific strain of *Verticillium lecanii* selected by GCRI for its high infectivity to aphids. This strain was isolated from a naturally diseased aphid in a commercial chrysanthemum glasshouse in southern England in the early 1970s. Despite considerable strain isolation and selection work since then, this strain remains the most infective to aphids of any tested. Under laboratory conditions this isolate has killed all aphid species tested, and in ideal conditions in the laboratory can give 100% infection of a population within 4–5 days.

The use of Vertalec for control of aphids on glasshouse chrysanthemums was developed by the UK Ministry of Agriculture, Fisheries and Food's Agricultural Development and Advisory Service, in particular by L.R. Wardlow of Wye, Kent, in association with the Glasshouse Crops Research Institute, Littlehampton.

Commercially, Vertalec is used to control aphids on glasshouse all-year-

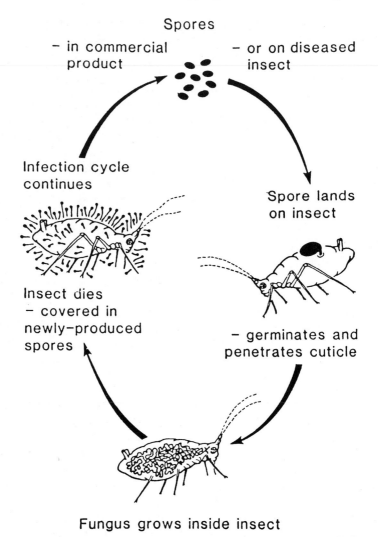

Fig. 2.1. *Verticillium lecanii* infection cycle (courtesy of Dr R.A. Hall).

round chrysanthemums, either grown in soil (for cut stems) or in pots. The most common pest species are *Aphis fabae*, *A. gossypii*, *Brachycaudus helichrysi*, *Macrosiphoniella sanborni* and *Myzus persicae*. In general, the product is more effective against the most mobile aphid species, probably owing to the greater spread of the disease among the population. Thus *Myzus persicae*, a highly mobile species, is more susceptible in the glasshouse than *Aphis gossypii*, which can escape control unless actually hit by

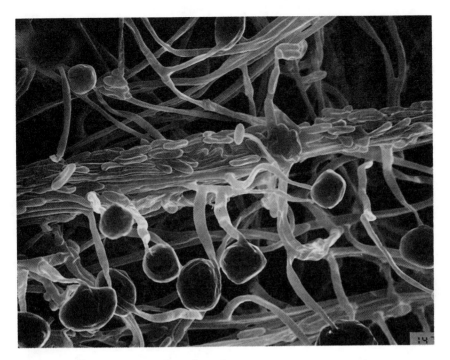

Fig. 2.2.   Hyphae of *Verticillium lecanii*.

the spore suspension. Similarly, aphids living in the more humid parts of the plant appear to be more susceptible than those, like *Macrosiphoniella sanbornii*, which live on the bare stems.

There are no examples of induced resistance to the fungus or indications that aphids resistant to chemical pesticides are any more or any less susceptible to the pathogen. Fungus from the body of one species remains infective to aphids of another species.

In the glasshouse, under suitable environmental conditions, *Verticillium lecanii* creates an epizootic within the aphid population that can last the length of the crop season. On the chrysanthemum crop, this is usually a period of 3 months. In a commercial glasshouse trial in Holland against the glasshouse whitefly *Trialeurodes vaporariorum* on cucumber, a single application of Mycotal was shown to control the insect for a period of over 4 months. Long-term control is a direct result of the ability of the fungus to grow on the insect cadaver and to subsequently infect healthy individuals to which viable spores adhere. This continuous control provides a clear market advantage over chemical pesticides, particularly the short-persistence contact insecticides, which may need to be applied at 10- to 14-day intervals.

The application procedure for both Mycotal and Vertalec begins by presoaking the formulation in water for 4–6 hours before application. This action first rehydrates the desiccated *Verticillium* spores and secondly aids the dispersion of the spores during application. Growers usually mix 500 g of product (the standard commercial pack size) with 2–3 litres of water in a clean bucket at midday in preparation for application in the afternoon.

Application is conventionally by high volume (either through a centralised spraying system or by knapsack), as it provides more water than would a low-volume system for germination and growth of the spores, as well as increasing local humidity and aiding good coverage. *Verticillium* is normally applied during the late afternoon or early evening in order to take advantage of the higher night-time humidity levels. The recommended application rate is 2.5 kg of product in 550–2000 litres of water per hectare, or for smaller areas 25 g in 6–20 litres/100 m^2. This equates to approximately 5×10^{11} viable colony-forming units of *Verticillium lecanii* per hectare. The spray is directed on to the undersides of the leaves and on to the growing points.

Despite *Verticillium*'s disease-like ability to spread within the insect population, all plants have to be treated and application to alternate rows is not recommended as the fungus does not spread fast enough to prevent plant damage or the build-up of excessive numbers on untreated plants.

2.3.3 *Vertalec on chrysanthemums*

Vertalec must be used early in the life of the chrysanthemum crop while aphid numbers are low, as a prophylactic treatment. Timing is critical for two reasons: late application will leave unsightly bodies on the crop, and the growth in the insect population may outstrip the infection rate. Vertalec is applied two weeks after planting out whether aphids are seen or not. If any survivors are seen by week 5, a second application should be made. After this date heavy infestations of more than 20 aphids per stem are normally treated with chemical aphicides.

These original recommendations for the use of Vertalec have been modified and adapted both by growers themselves and by the Ministry of Agriculture (ADAS) advisors to suit local practice and to overcome adverse (particularly low humidity) conditions. As a result, in addition to the single application system described, an alternative low dose/low volume, multiple application system has been developed. One version of the multiple dose system (and that recommended by ADAS) uses Vertalec at 250 g/ha (equivalent to 0.5 g per 1000 plants) in as little water as practicable applied twice weekly throughout the life of the crop. Other recommendations pertaining to successful use of *V. lecanii* remain, so that, for example, the application should take place in the late afternoon or early evening.

One method that has been developed for the multiple application system has been use of overhead irrigation lines and diluter systems. This has proved easy and effective.

The 'low and often' approach to the use of Vertalec has been shown to be more effective against some of the more sedentary aphid species than the 'high dose–one dose' system, the deficiency in the one dose method commonly being that the disease does not spread fast enough or efficiently enough to successfully control such aphid species (especially *Aphis gossypii*) in sub-optimal environmental conditions. The multiple dose system artificially promotes disease spread and hence has proved to be more reliable against that insect or in otherwise marginal, low-humidity glasshouses.

2.3.4 *Practical factors for successful use*

The successful performance of *V. lecanii* within the glasshouse is totally dependent on the suitability of the temperature and relative humidity conditions within the crop cover, and also on timing of application. Precise understanding of these conditions is still inadequate and cannot easily be summarised in terms of average conditions or even extremes. General recommendations have therefore been difficult to make. However, the following conditions have been described as the minimum required for successful use.

Relative humidity: Never below 60% rh.
Normal night-time above 85% rh.
Temperature: Never above 30°C or below 10°C.
Normal night-time between 15 and 25°C.

Under normal crop conditions, without supplementary heating, the diurnal variation in these conditions will ensure the availability of sufficient free water for infection. However, if the humidity in the crop is abnormally low during a period of settled fine weather in midsummer, then the presence of free water can be achieved by light damping down.

Extensive studies have been carried out on the compatibility of *V. lecanii* with various other pest control agents commonly used within the glasshouse. This is necessary so that compounds deleterious to the fungus can be avoided if possible. Similarly, the possible impact of *V. lecanii* on the beneficial organisms being used within the glasshouse to control other pests must also be known. In practice, fungicide use is the prime concern when either Vertalec or Mycotal have been or are being applied. Many of the standard broad-spectrum compounds used by growers, such as benomyl, captan or thiram, are very toxic to the fungus and cannot be sprayed at the same time as *Verticillium*. Table 2.2 lists compatible and incompatible compounds, from which growers can select materials which are the least problematic. However, even highly inhibitory compounds like benomyl

Table 2.2.   Compatibility of *Verticillium lecanii* with chemical pesticides

---

Group A: Compatible/tank mix   Mycotal and Vertalec can be applied as a tank mix with or on the same day as the following chemicals:
    Fungicides: benodanil; carbendazim; etradiazole; oxycarboxin; pyracarbolid; quintozene; vinclozolin
    Insecticides: bioresmethrin; deltamethrin; dichlorvos; dienochlor; diflubenzuron; endosulfan; HCH; permethrin; pirimicarb; white oil

Group B: Compatible/do not tank mix   Mycotal and Vertalec can be used but not tank-mixed or sprayed on the same day as the following compounds:
    Fungicides: bupirimate; dinocap; dimethirimol; pyrazophos; triadimefon; triforine
    Insecticides: carbaryl; cyhexatin; dicofol; dioxathion; heptenophos; methidathion; oxydemeton-methyl; tetradifon

Group C: Incompatible   Mycotal and Vertalec are not compatible with the following compounds, the application of which will kill the fungus. If use of these compounds is essential or unavoidable, users should integrate the chemical within a spray programme or a repeated application of Mycotal and Vertalec a week after application of the fungicide. It is normally advisable to utilise one of the less harmful compounds in Groups A or B if possible.

    Fungicides: benomyl; captan; chlorothalonil; dichlofluanid; fenarimol; fentin acetate; folpet; imazalil; iprodione; mancozeb; maneb; nithrothal-isopropyl/zinc of sulphur; sulphur; thiram
    Acaricides: propargite

---

can be used without loss of infectivity if applied a week before *V. lecanii* is applied or a week after (Gardner *et al.*, 1984).

Both *Verticillium* products are compatible with other commercially available biological control agents (such as the spider mite predator *Phytoseiulus persimilis* or the caterpillar bacterium *Bacillus thuringiensis*), and have no adverse effect on their activity. *V. lecanii* will infect the whitefly parasite *Encarsia formosa*, but does not appear to affect the performance of the parasite population as a whole in the glasshouse.

Despite the specific recommendations for use, backed up by careful monitoring, the key problem in obtaining control with the *Verticillium* products is persuading the user to read the instructions and to take notice of them. Thus, in its first year of commercial use, 1981, Vertalec was sometimes tank-mixed with benomyl, used as a clean-up spray two days before harvest, expected to kill a wide range of insect pests, stored in the glasshouse (at temperatures up to 35°C), expected to kill insects instantly, used outdoors, applied early in the morning rather than late in the evening (causing the inoculum to become desiccated), sprayed at fractions of the recommended rate, and applied to fractions of the crop (on the assumption that it would spread from one part of the glasshouse to another fast enough to give adequate control)!

Each of the problems described would have led to inadequate control and hence to the product being branded by the user as non-effective. The glasshouse industry is fairly close-knit, and reputation for, or rumour of, failure spreads quickly. The introduction of fungal entomopathogens therefore needs not only a thorough understanding of the pest and pathogen, but also a well developed technical procedure for use, and furthermore a very careful and structured method for product introduction. Explanation to the grower is all important, and mistakes in this area may reduce or slow down the market acceptance of this type of product.

2.4 The future

Although a large volume of literature has been published to date on the biology and use of entomopathogenic fungi, and it is growing yearly, the development of these organisms as pest control agents is still in its infancy. In particular, a better understanding is needed of strain selection and manipulation, and especially of the infection process. In essence, we still lack basic knowledge on what makes a 'good' fungal isolate and on what features in a 'good' fungal entomopathogen combine to make it into a successful insecticide.

Early work at Kings College, London, begun in association with Microbial Resources and the Glasshouse Crops Research Institute, is beginning to make some important laboratory breathroughs in this area (see Heale, Chapter 11 of this volume). This work will have to be supported by more elaborate glasshouse studies, where pest and fungus biology as well as environmental factors will have to be determined and their implications understood. A continuing complication is that the picture is periodically altered by changes in growing practice, pest occurrence and pesticide use.

Only with a greater understanding of the biology of the fungus and its interaction with the host crop plant, the other pests and control agents in use, and the environmental conditions, will further advances in the use of these fungi occur. Researchers can then target their activity to overcome some of the problems related to fungal pest-control agents.

The key areas for development of entomopathogenic fungi are therefore:

(1) Improved tolerance of low humidity.
(2) Improved infection at low humidity.
(3) Improved growth at low temperature.
(4) Improved dispersal (epizootic).
(5) Faster *in situ* growth and hence faster kill.
(6) Greater (broader) fungicide tolerance.
(7) Improved product shelf-life.

(8) Reduced cost by cheaper production (better productivity) and increased infectivity.

In essence, these factors can be summarised as more reliable potency, under less stringent environmental conditions, at lower cost. Until these criteria are met, fungal pathogens in temperate climates will not only be restricted to glasshouses, but will largely continue to be of only limited interest there. However, such 'improvements' should result in considerable growth in the use of these organisms. Fungi have the potential to provide the grower with a safe, self-sustaining and selective pesticide, with little likelihood of induced pest resistance; a combination of features which should be highly desirable to anyone in the glasshouse industry with an insect pest problem. In addition, knowledge gleaned from use of fungi in glasshouses should be useful in applying them outside, particularly in the humid tropics and sub-tropics.

Acknowledgements

The author would like to thank Dr Denis Burges for reading the text and providing helpful comments.

References

Adam, Kh. (1983). Results and aims of the simultaneous introduction of the parasite *Encarsia* and the entomopathogenic fungus *Aschersonia* for the biological control of the glasshouse whitefly on cucumber in glasshouses. *Informatsionnyi Byulleten VPS MOBB*, **9**, 50–2.

Borisov, B.A. & Vinokurova, T.P. (1983). Increasing the effectiveness of entomopathogenic fungi. *Zashchita Rastenii*, **9**, 20–2.

Brandenburg, R.L. & Kennedy, G.G. (1983). Interactive effects of selected pesticides on the two-spotted spider mite and its fungal pathogen *Neozygites floridana*. *Entomologia experientia et applicata*, **34**, 240–4.

Gardner, W.A., Oetting, R.D. & Storey, G.K. (1982). Susceptibility of the two-spotted spider mite, *Tetranychus urticae* Koch, to the fungal pathogen *Hirsutella thompsonii* Fisher. *Florida Entomologist*, **65**, 458–65.

Gardner, W.A., Oetting, R.D. & Storey, G.K. (1984). Scheduling of *Verticillium lecanii* and benomyl applications to maintain aphid (Homoptera: Aphidae) control on chrysanthemums in greenhouses. *Journal of Economic Entomology*, **77**, 514–18.

Gillespie, A.T., Hall, R.A. & Burges, H.D. (1982a). Control of onion thrips, *Thrips tabaci* and the red spider mite, *Tetranychus urticae* by *Verticillium lecanii*. *Proceedings of the IIIrd International Colloquium of Invertebrate Pathology*, Papers, 101.

Gillespie, A.T., Hall, R.A. & Burges, H.D. (1982b). Entomogenous fungi as control agents for the glasshouse leafhopper, *Zygina pallidifrons*. *Proceedings of the IIIrd International Colloquium of Invertebrate Pathology*, Papers, 108.

Hall, R.A. (1981). The fungus *Verticillium lecanii* as a microbial insecticide against aphids and scales. In: *Microbial Control of Pests and Plant Diseases 1970–80*, ed. H.D. Burges, pp. 483–98. Academic Press, London.

Hall, R.A. (1982a). Control of whitefly *Trialeurodes vaporariorum* and cotton aphid *Aphis gossypii* in glasshouses by two isolates of the fungus *Verticillium lecanii*. *Annals of Applied Biology*, **101**, 1–12.

Hall, R.A. (1982b). The use of *Verticillium lecanii* to control whitefly and other pests. *Proceedings of the IIIrd International Colloquium of Invertebrate Pathology*, Papers, 95.

Hall, R.A. & Burges, H.D. (1979). Control of aphids in glasshouses with the fungus *Verticillium lecanii*. *Annals of Applied Biology*, **93**, 235–46.

Ignoffo, C.M. (1981). The fungus *Nomuraea rileyi* as a microbial insecticide. In: *Microbial Control of Pests and Plant Diseases 1970–80*, ed. H.D. Burges, pp. 513–38. Academic Press, London.

Keller, S. & Wuest, J. (1983). Observations on three species of *Neozygites* (Zygomycetes: Entomophthoraceae). *Entomophaga*, **28**, 123–4.

Kitizawa, K., Fujisawa, I. & Imabayashi, S.I. (1984). Isolation of *Verticillium lecanii* affecting aphids and greenhouse whitefly in Japan. *Annals of the Phytopathological Society of Japan*, **50**, 574–81.

Landa, Z. (1984). Protection against glasshouse whitefly (*Trialeurodes vaporariorum* Westw.) in integrated protection programmes for glasshouse cucmbers. *Sbornik UVTIZ, Zahradnictvi* **11**, 215–28.

Marletto, O.O. & Maggiora, S. (1983). Deuteromycete parasites of *Zyginidia pullula*. *Atti XIII Congresso Nazionale Italiano di Entomologia*, Turin, Italy, 539–44.

Morgan, W.M. & Ledieu, M.S. (1979). Pests and diseases of glasshouse crops. In: *Pest and Disease Control Handbook*, eds N. Scopes & M. Ledieu, pp. 7.1–7.81. BCPC, Croydon.

Quinlan, R.J. (1982). The control of glasshouse whitefly on tomatoes using *Verticillium lecanii*. *Proceedings of the IIIrd International Colloquium of Invertebrate Pathology*, Papers, 230.

Quinlan, R.J. (1983). The use of *Verticillium lecanii*, an entomopathogenic fungus, to control glasshouse whitefly (*Trialeurodes vaporariorum*). *Proceedings of the 10th International Congress of Plant Protection*, **2**, 787.

Ramakers, P.M.C., Rombach, M.C. & Samson, R.A. (1982). Application of the entomopathogenic fungus *Aschersonia aleyrodis* in an integrated control programme against the glasshouse whitefly *Trialeurodes vaporariorum*. *Proceedings of the IIIrd International Colloquium of Invertebrate Pathology*, Papers, 99.

Roberts, D.W. (1981). Toxins of entomopathogenic fungi. In: *Microbial Control of Pests and Plant Diseases 1970–80*, ed. H.D. Burges, pp. 441–64. Academic Press, London.

Samson, R.A. (1981). Identification: entomopathogenic deuteromycetes. In: *Microbial Control of Pests and Plant Diseases 1970–80*, ed. H.D. Burges, pp. 93–106 Academic Press, London.

Samson, R.A., Ramakers, P.M.J. & Oswald, T. (1979). *Entomophthora thripidum*, a new fungal pathogen of *Thrips tabaci*. *Canadian Journal of Botany*, **57**, 1317–23.

Soares, G.G. Jr, Marchal, M. & Ferron, P. (1983). Susceptibility of *Otiorhynchus sulcatus* (Coleoptera: Curculionidae) larvae to *Metarhizium anisopliae* and *Metarhizium flavoviride* (Deuteromycotina: Hyphomycetes) at two different temperatures. *Environmental Entomology*, **12**, 1886–1890.

Treifi, A.H. (1984). Use of *Beauveria bassinana* (Bals) to control the immature stages of the whitefly *Trialeurodes vaporariorum* (Westw.) (Homoptera: Aleyrodidae) in the greenhouse. *Arab Journal of Plant Protection*, **2**, 83–6.

Wilding, N. (1981). Pest control of Entomophthorales. In: *Microbial Control of Pests and Plant Diseases 1970–80*, ed. H.D. Burges, pp. 539–54. Academic Press, London.

Wilding, N. (1983). The current status and potential of entomogenous fungi as agents of pest control. *Proceedings of the 10th International Congress of Plant Protection*, **2**, 743–50.

Zimmerman, G. (1984). Further trials with *Metarhizium anisopliae* (Fungi Imperfecti, Moniliales) for control of the black vine weevil, *Otiorhynchus sulcatus* F., on pot plants in the greenhouse. *Nachrichtenblatt des Deutschen Pflanzenschutzdienstes*, **36**, 55–9.

3 *A.T. Gillespie*

Use of fungi to control pests of agricultural importance

3.1 Introduction

The potential of entomogenous fungi for pest control has been realised since the 19th century when Metchnikoff (1879) and Krassilstschik (1888) mass-produced *Metarhizium anisopliae* and tested preparations for control of the wheat cockchafer, *Anisoplia austriaca*, and sugarbeet curculionid, *Cleonus punctiventris*. At about the same time large-scale field experiments with fungi were conducted in Europe (see Ignoffo, 1981). Early in the 20th century Berger (1909) worked extensively on *Aschersonia* in the United States, and this fungus was produced until the 1920s when effective chemical insecticides became available. It was concluded that successful use of the fungus was too dependent on high relative humidity. Since that time there have been many attempts to use fungi to control agricultural pests but very few successes. Indeed, products have been launched commercially and then withdrawn. At present, there is no commercial production of fungi for pest control in the Western world though fungi are used on a moderate scale in China, eastern Europe and South America.

This chapter examines attempts to develop fungi for pest control. Particular attention is devoted to those areas where major developments are required before the use of fungi can be adopted on a large scale; ideas for such improvements are also presented.

3.2 Taxonomic classification of entomopathogenic fungi

Entomogenous fungi occur primarily in the Deuteromycetes and Entomophthoraceae. Genera can be highly specific in their host range, e.g. the deuteromycete *Aschersonia* infects only scale insects and whitefly, and *Massospora* (Entomophthoraceae) only cicadas, whereas other fungi have been recorded from many orders of insects, e.g. *Beauveria* and *Metarhizium* (Deuteromycetes) and *Zoophthora radicans* (Entomophthoraceae).

Some Deuteromycetes are now known to have teleomorphs (sexual stages); e.g. a *Torrubiella* sp. was recently described as the teleomorph of *Verticillium lecanii* (Evans & Samson, 1986). The Entomophthoraceae comprise a number of genera, which occur mainly in temperate and continental areas and are characterised by their ability to discharge conidiospores forcibly. This ability enhances their potential to spread and cause epizootics. The taxonomy of the Entomophthoraceae has been the subject of much debate in recent years, sometimes resulting in conflicting classifications. However, the classification of Remaudière & Keller (1980) is now generally accepted, with the Entomophthoraceae being divided into the genera *Conidiobolus, Entomophthora, Erynia, Massospora, Neozygites* and *Zoophthora*. It should be noted that a modified classification has been proposed which additionally includes the genera *Entomophaga, Strongwellsea* and *Tarichium* (Humber, 1981; Ben Ze'ev & Kenneth, 1982a, b).

3.3 Biology

The infective unit of entomogenous fungi is the conidium. Deuteromycetes form only primary conidia, whereas Entomophthoraceae form primary, secondary, tertiary or even quaternary infective conidia. At favourable temperatures and in the presence of sufficient water, conidia germinate, generally on the host cuticle, and form a germ tube. This either penetrates the cuticle directly, as in *Verticillium lecanii* (Hughes & Gillespie, 1985) and *Beauveria bassiana* (U. Hilber & A.T. Gillespie, unpublished observation), or forms an appressorium which produces a penetrant peg, e.g. *M. anisopliae* (Zacharuk, 1970 a,b,c).

For Deuteromycetes, the critical water activity (a_w) for germination and mycelial growth is about 0.92 (equivalent to a relative humidity (r.h.) of 92%. Thus, conidia of *B. bassiana*, *M. anisopliae*, *Paecilomyces* spp. and *V. lecanii* all germinate at 0.94 a_w but not at 0.92 a_w (Gillespie & Crawford, 1986). Furthermore, rates of germination and mycelial growth were maximal at 0.99, 0.98 and 0.97 a_w and were markedly reduced at 0.96 and 0.94 a_w. In one isolate of *M. anisopliae*, growth was morphologically abnormal at 0.96 a_w. Similar data demonstrating the importance of moisture to aphid infection by *V. lecanii* were obtained with aphids living on leaves (Milner & Lutton, 1986). Disease transmission was maximal in the presence of free water, less at 100% r.h., still less at 97% r.h. and absent at 93% r.h. Clearly, water availability is of vital importance for fungal growth, and small differences in the relative humidity levels occurring after spore application can determine whether or not the fungus can successfully control pests.

Penetration of insect cuticle by germinating spores has long been thought to result from a combination of enzymic degradation of the cuticle and mechanical pressure by the germ tube. Strains of *B. bassiana*, *M. anisopliae*, *Paecilomyces* spp. and *V. lecanii* all produced large quantities of proteases and chitinases in liquid culture (St. Leger *et al.*, 1986a). Production of protease, lipase and chitinase on insect cuticle has also been demonstrated with *M. anisopliae* by enzyme-specific staining and recovery of enzymes from fly wings previously inoculated with conidia (St. Leger *et al.*, 1987). In several of *B. bassiana* and *M. anisopliae* strains, the key enzyme is an endoprotease which dissolves the protein matrix masking cuticular chitin (Smith *et al.*, 1981; St. Leger *et al.*, 1986b). Chitinase production occurs later in the infection process.

Once through the cuticle, the fungus must overcome the host defence system before it can enter the haemolymph and spread throughout the insect. In some fungi, toxins (see review by Roberts, 1981) may inactivate the host defence system. A crude, partially purified extract from a *M. anisopliae* culture filtrate containing the cyclic depsipeptides, destruxins, inhibited prophenol oxidase production by insect haemocytes, suggesting that destruxins may suppress insect immune responses (Huxham *et al.*, 1986). Once in the haemolymph, growth morphology changes and the fungus assumes a 'yeast-like' form which multiplies by budding. The fungus spreads through the haemocoel, killing the insect 3–14 days after spore application, with the precise time dependent on spore dosage and on temperature. Probably death eventually results from a combination of mechanical damage (causing water loss), nutrient exhaustion and toxicosis.

After death, provided water availability is high, the fungus emerges outwards through the cuticle and sporulates on the cadaver, providing inoculum for infection of further insects. If conditions are unfavourable, the fungus remains inside the insect, where it can survive for several months, eventually producing spores when favourable conditions return. *Erynia neoaphidis* survives inside aphid cadavers for 32 weeks at 0°C and 20% or 50% r.h., also for 8 weeks at 20°C and 20% r.h. (Wilding, 1973). Such mycelial persistence in cadavers is likely to be an important mechanism of fungal survival, particularly in those species unable to produce resting spores. Several species of the Entomophthoraceae produce thick-walled resting spores, which survive adverse conditions, e.g. resting spores of *Conidiobolus obscurus* survived for 12 months at 6°C (Perry *et al.*, 1982). The switch from infective conidia to resting spores is perhaps stimulated in some species by environmental trigger factors such as changes in daylength. Thus the thrips pathogen, *Neozygites parvispora*, produces mainly conidia early in the season, but primarily resting spores in the autumn (Carl, 1975). Deuteromycetes probably overwinter as conidia or mycelia in cadavers since specialised resting spores are generally absent, though thick-walled resting spores in *M. anisopliae* were reported by Zacharuk (in Ferron, 1981). Conidia can survive in soil for long periods: *B. bassiana* conidia have a half-life of 276 days at 10°C (Lingg & Donaldson, 1981) and some *M. anisopliae* conidia survive for at least 21 months at 19°C (Fargues & Robert, 1985).

3.4 Production of fungi

Usually, fungi have been used as mycoinsecticides, involving large amounts of fungal inoculum. Thus, ease of maintaining cultures and of producing fungi is of great importance.

Entomophthoraceae are frequently difficult to grow in axenic culture and require sophisticated media: some are obligate parasites, e.g. *N. parvispora* (Carl, 1975). Because their conidia are thin-walled and survive poorly, they are unsuitable to harvest for pest control. In contrast, resting spores store well and can be produced in liquid fermentation in some species. Such a method was developed for *Conidiobolus obscurus*, with yields of 2×10^6 resting spores per millilitre (Latgé *et al.*, 1978). However, these resting spores failed to provide aphid control in the field (see below). Recently, McCabe & Soper (1985) described a process for preserving fermenter-produced mycelium. The fungal mycelium is dried and ground to a coarse powder, which is rehydrated and sprayed in the field, where it produces infective conidia. The method has not yet been commercially exploited. It has the disadvantages that the dried mycelium stores poorly at moderate temperatures and that the time required to produce infective spores reduces the speed of insect kill.

Deuteromycete spores are often easier to produce and store than those of the Entomophthoraceae. Conidia of many genera can be produced in large quantities on semi-solid media, e.g. hydrated cereal grains and oil (Aquino *et al.*, 1975), or even in submerged culture, e.g. *B. bassiana* (Goral, 1979; Thomas *et al.*, 1987) and *Hirsutella thompsonii* (Van Winkelhoff & McCoy, 1984). However, in submerged culture, most fungal strains produce hyphal bodies (blastospores) similar to those formed in insect haemolymph during fungal infection. Strains of *B. bassiana*, *Beauveria brongniartii* or *V. lecanii* readily produce more than 10^{12} spores per litre of fermenter medium. Several methods of processing these spores for storage have been devised, e.g. silica powder was used to dry *B. brongniartii* spores (Blachère *et al.*, 1973) and the effect of clay coating on survival of *B. bassiana* spores was studied by Fargues *et al.* (1983).

Further details of production and formulation of fungi as mycoherbicides is provided by Bartlett & Jaronski (Chapter 4).

3.5 An example of classical biological control

After introduction into Australia during the late 1970s, the lucerne aphid, *Therioaphis trifolii*, rapidly became a serious pest of lucerne. A search for pathogenic fungi in Australia was unsuccessful, but several isolates of *Zoophthora radicans* from Israel were found to be pathogenic. After appropriate testing and introduction into Australia, the exotic *Z. radicans* isolate rapidly became established and spread (Milner *et al.*, 1982). The fungus now partially controls the lucerne aphid and saves many millions of dollars annually in insecticide costs (R.J. Milner, personal communication). The aphid became a pest because it had been introduced into Australia without its natural enemies. The introduction strategy with the fungus worked because the environment was sufficiently moist to provide a niche for the introduced fungus. This classical biological control approach has not often been adopted, though its potential is probably great.

3.6 Entomophthoraceae: problems and potential

In temperate agricultural ecosystems species of Entomophthoraceae contribute greatly to maintaing pest numbers below economic damage thresholds. Thus, a study of parasites, predators and pathogens occurring on potato aphids showed that *Erynia neoaphidis*, *C. obscurus* and *Z. radicans* were the most important natural factors in maintaining low aphid numbers (Shands *et al.*, 1962). Similarly, several species of Entomophthoraceae are important natural agents in maintaining low populations of aphids on cereals (Remaudière *et al.*, 1981). In the light of these ecological studies it is surprising that few workers have conducted field experiments. Wilding

(1981, 1982) reviewed the literature and recorded a few instances of successfully applied field control. Application of an aqueous spray containing homogenised larvae killed by *Z. radicans* controlled *Plutella xylostella* on brassicas (Kelsey, 1965). Suspensions of *C. obscurus* conidia, hyphal bodies and resting spores, produced *in vitro*, killed 74% of apple aphids without affecting natural predators (Cinovskis *et al.*, 1974). Wilding (1981) also reports the use of *E. neoaphidis* to control aphids in China. Field releases of inoculated insects have also provided successful control: release of *Aphis fabae* infected with *E. neoaphidis* and *Neozygites fresenii* into aphid populations on field beans resulted in a maximum 48% infection of apterous aphids, with a resultant doubling of bean yields compared with the control plot (Wilding, 1982). However, fungal treatments were inferior to insecticide applications, and good control could not be obtained consistently. Moreover, the application method was laborious and uneconomic. A more attractive application method was attempted utilising resting spores of *C. obscurus* produced by liquid fermentation (Latgé *et al.*, 1977). Spores were tested at rates of $5-20 \times 10^{10}$ per hectare for control of cereal and pea aphids during 1980. Aphid numbers were unaffected at all sites (Keller, 1981; Remaudière, 1981). Extrapolating from laboratory data, Latgé (1982) calculated that $1-2 \times 10^{12}$ resting spores per hectare would be necessary to provide LD_{50} doses in the field which would require the production of 1000–2000 litres of fermenter medium. As Latgé pointed out, this approach is totally unrealistic. Later, Latgé (1986) listed possible reasons for failure of insect control with Entomophthoraceae: (a) the most pathogenic species do not produce resting spores; (b) the resting spores are not the infective propagule; and (c) the production of resting spores is uneconomic, owing to the low yields and the lengthy production process needed for spore maturation and breaking of dormancy. Because of the disappointing results obtained with resting spores, recent effort has been devoted to production of hyphal bodies. These propagules are easily produced in liquid culture but have a short shelf-life, are not themselves infective and require favourable conditions after application to promote sporulation. Recent developments have been made in formulating hyphal bodies and mycelia (McCabe & Soper, 1985), but further formulation developments to improve storage and lessen reliance on high relative humidity after application are essential if the use of Entomophthoraceae for pest control is to become a reality.

3.7 Aschersonia: an early example of mycoinsecticide use

Early in the 20th century, *Aschersonia aleyrodis* was used extensively in the humid citrus groves of Florida to control scales and whitefly, and other fungi — e.g. *Sphaerostilbe coccophila* — were used for control of the San

José scale (Rolfs & Fawcett, 1908). Several enterprising growers collected fungus-infected insects and sold them to citrus growers at a cost of $2–3 per acre. Researchers cultured *A. aleyrodis* on sweet potato and produced spores for application to pest-infested trees. Even at this time workers were well aware of the fungi's requirements for high relative humidity. The annual report of the Florida Agricultural Station for 1909 described what happened after an application of *Aschersonia* spores: 'The weather being dry, the fungus was slow in starting to develop, until the drought was broken.' The use of *A. aleyrodis* ended when effective chemicals became available in the 1920s. More recently, *A. aleyrodis* has been used for whitefly control in glasshouses in The Netherlands and eastern Europe, with some success.

3.8 Beauveria bassiana: a mycoinsecticide for field and forest

Beauveria bassiana is perhaps the most extensively studied entomogenous fungus and has been isolated from a variety of insects worldwide. This research resulted in moderate use of *B. bassiana* in the People's Republic of China and in eastern Europe. In China, *B. bassiana* is produced in communes and is applied against the European corn borer, *Ostrinia nubilalis*, pine caterpillars, *Dendrolimus* spp., and green leafhoppers, *Nephotettix* spp. The area treated with *B. bassiana* in 1977 was quoted as several 100 000 ha (Franz & Krieg, 1980). Another estimate reports that 400 000 ha were treated for corn borer alone (Hussey & Tinsley, 1981). Spores are produced on boiled rice contained in pits in the ground and covered after inoculation with a liquid culture of *B. bassiana* (D.E. Pinnock, personal communication). The semi-solid fermentation presumably excludes large-scale contamination because *B. bassiana* produces potent antibiotics, e.g. oosporein.

In the Soviet Union, *B. bassiana* was produced under the trade name Boverin for control of the Colorado potato beetle, *Leptinotarsa decemlineata* and the codling moth, *Laspeyresia pomonella* (Ferron, 1981). Fungi caused considerable natural mortality of *L. decemlineata* with, on occasions, 80–100% of over wintering beetles succumbing to fungal disease (Sikura & Sikura, 1983). Boverin contained 2×10^9 conidia per gram and was stored for up to 4 months at temperatures from 18 to 22°C (Lipa, 1985). *B. bassiana* more readiy infects insects that are physiologically weakened, and in practice Boverin was used in combination with insecticides. Without insecticides, 3 or 20 kg per hectare of Boverin were required to kill 90% of 1st or 4th instar larvae, respectively. Obviously, such high doses were uneconomic, and 2 kg of Boverin were recommended together with a quarter dose of insecticide (400 g), normally trichlorphon, per hectare. It is recommended that the treatment be applied twice (Kravtsov,

1982). Boverin was used on 7900 ha in 1977 (Lipa, 1985). Also attempts have been made to control Colorado beetle with *B. bassiana* in the US. At least one company has produced a pilot product and tested it extensively in field trials. The results have been variable, with generally poor control being obtained, but occasionally spore application has resulted in spectacular epizootics (Watt & LeBrun, 1984). This lack of reliability limits commercial development of this fungus for control of Colorado potato beetle.

However, the prospects for developing *B. bassiana* to control certain pests appears promising. Conidial application to young citrus trees provided protection from weevils for up to 6 months (McCoy, in Roberts and Wraight, 1986). *B. bassiana* has also shown great potential for control of the European corn borer, *Ostrinia nubilalis*. Riba (1984) described some 80% mortality of borers after placing conidia in the whorl of young corn plants, and went on to demonstrate the reliability of *B. bassiana* in controlling this pest: 18 experiments were conducted in France during 1985, and in 17 cases *B. bassiana* performed as well as chemical insecticides (in Roberts and Wraight, 1986). *B. bassiana* was also moderately pathogenic to pupae of the onion fly *Delia antiqua*, though *Paecilomyces farinosus* and *P. fumoso-roseus* showed greater activity (Poprawski *et al.*, 1985).

From the available reports it seems that *B. bassiana* has a future as a mycoinsecticide but only in certain well-defined situations where applied spores are protected in an environment with favourable humidity.

3.9 Beauveria brongniartii: a cockchafer pathogen

Beauveria brongniartii has been extensively studied for control of the May beetle or European cockchafer, *Melolontha melolontha*, in Switzerland. Larvae of this pest are polyphagous feeders on the roots of most cultivated plants, including trees.

During the 20th century many attempts have been made to control *M. melolontha* with fungi of various genera. However, research resulted in the conclusion that *B. brongniartii* should be considered the most promising of these (Hurpin & Robert, 1972). Soil incorporation of *B. brongniartii* blastospores at 2×10^{13} and 2×10^{14} spores per hectare provided some control of larvae. In untreated plots, mortality ranged from 0 to 32% and from 14.5 to 56.6% at the lowest dose, compared with 38–80% with the highest application rate (Ferron, 1978).

Keller (1978) reported application of spores to adult female cockchafers and subsequent mortality in their larvae, thus demonstrating spread of the disease to progeny. In 1976, blastospores were applied along the border of a small forest to a swarming population, at 2.5×10^{14} spores per hectare. In subsequent years the population density was determined and samples of collected larvae were maintained in the laboratory. Field populations were similar in treated and untreated plots, and a maximum mortality of only 32% occurred in larvae quarantined in the laboratory (Keller, 1979). These disappointing data led to the conclusion that control of *M. melolontha* was not possible with *B. brongniartii*. However, field populations treated with the fungus collapsed during the second generation, encouraging the continuation of the programme (Keller, 1983). In 1985, 14 swarming sites, totalling 89 ha of forest, were treated at a rate of 2.6×10^{14} blastospores per hectare, which caused mean larval infection rates of 85% (Keller, 1986). In the experiment, spore application rates were high. Assuming a spore yield of approximately 10^{12} per litre, the product of 250 litres of fermentation broth would be required to treat each hectare, at a basic production cost of about £25. When drying, formulation, packaging and distribution costs are included for the potential product, this high rate treatment becomes very expensive. However, the cost would be acceptable because prolonged pest control is obtained over several years, and several thousand hectares are protected by treating swarming sites. Obviously, the use of isolates of greater virulence and/or increased spore yields will help to improve the economics still further and increase the likelihood of this pest control method being adopted. It should be noted that similar conclusions concerning the economics of cockchafer control by *B. brongniartii* were reached by Ferron (1981).

3.10 **Metarhizium anisopliae: a versatile mycoinsecticide**

Metarhizium anisopliae occurs as two forms differing in conidial size (Tulloch, 1976). *M. anisopliae* var. *anisopliae* has been isolated from many insects while var. *major* occurs only on the rhinoceros beetle, *Oryctes rhinoceros*. *M. anisopliae* is used in Brazil to control spittlebugs (Homoptera: Cercopidae), e.g. *Mahanarva posticata* on sugar cane. In 1978, 50 000 ha of sugar were treated in the state of Pernambuco alone. Aerial sprays at the low rate of 6×10^{11}–1.2×10^{12} conidia per hectare caused a maximum mortality of 65%, which was a sufficient reduction to provide increases in cane sugar content (Ferron, 1981). The fungus is produced by sugar growers' co-operatives on a substrate of sterilised rice contained in glass bottles or autoclavable polyethylene bags (Aquino *et al.*, 1975). The cereal grains and spores are dried, ground and distributed under a variety of trade names, e.g. Metaquino and Metabiol. Yields average 10^9 conidia per gram of rice, which means 1 kg of rice produces sufficient spores to treat a hectare of sugar. The process is therefore very inexpensive, and possibly higher doses of conidia could be used to increase insect mortality.

  M. anisopliae is also being developed for control of the pasture cockchafer, *Aphodius tasmaniae*, in Australia. Populations can reach 2000 larvae per square metre, causing severe loss of autumn and winter pasture. *M. anisopliae* conidia were produced on bran and the fungus was applied as a bait at rates of 1.4×10^{14} or $1.3–3.0 \times 10^{15}$ spores per hectare (Coles & Pinnock, 1984). Four months after application about 80% mortality had occurred at each site, with 55% occurring at one site after only one month. The fungus persisted to a limited extent and one year after application between 12 and 20% of larvae were infected with *M. anisopliae*. In these experiments the spore dose was high, but recent studies have improved efficacy (primarily by improvements in bait formulation) and control is now more economic (D.E. Pinnock, personal communication).

Termites are a serious problem in many areas of the world and *M. anisopliae* is a potential control agent. The termite *Nasutitermes exitiosus* was susceptible to *M. anisopliae* in the laboratory (Hänel, 1982). Field experiments indicated that contamination of small numbers of termites resulted in the spread of fungal spores to large numbers of insects. Some colonies were eliminated whereas other populations declined only slowly even though insects were contaminated with sufficient conidia to provide mortality under conditions which laboratory tests had shown to be optimal for the fungus. Further studies are needed to determine the full potential of *M. anisopliae* for termite control.

M. anisopliae var *major* is used in parts of the Pacific and south-east Asia to help control the serious coconut pest *O. rhinoceros*; it is applied to give rapid control in certain moist situations where larval densities are high to supplement the main method of control, i.e. a baculovirus, which spreads effectively over an extended period of time. Generally, conidia are produced on rice and the cereal and fungus distributed in known beetle breeding sites (Bedford, 1980).

3.11 Hirsutella spp: pathogens in the tropics

Most species of *Hirsutella* have been isolated from diseased insects and mites in the tropics. Some are important pathogens of agricultural pests, e.g. *Hirsutella citriformis* commonly causes epizootics on the rice brown planthopper, *Nilaparvata lugens* (R.A. Hall, unpublished observation) and *Hirsutella versicolor* infects mango jassids (A.T. Gillespie, unpublished observation). However, most *Hirsutella* species have received little study, perhaps because many are difficult to grow in culture or they sporulate poorly. The exception is the specific mite pathogen, *Hirsutella thompsonii*, which has been intensively studied for control of the citrus rust mite, *Phyllocoptruta oleivora*.

Under optimal conditions of temperature and humidity, *H. thompsonii* conidia can penetrate the integument within 4 hours and within 72 hours the host dies and fungal sporulation commences (McCoy & Selhime, 1974).

Under certain conditions, thick-walled chlamydospores are formed inside infected mites and allow fungal survival during adverse conditions (McCoy, 1981). *H. thompsonii* commonly causes epizootics in *P. oleivora* in Florida that virtually eliminate mites on citrus (McCoy, 1981), but too late to prevent mites damaging trees. In many field experiments, *H. thompsonii* has been applied before crop damage occurred. Thus laboratory-produced mycelia, applied to moderate to high mite populations on mature citrus trees, reduced mite populations to low levels within 2 weeks and maintained control for 10–14 weeks (McCoy *et al.*, 1971). The applied mycelia were not able to infect insects directly but sporulated to produce infective conidia. Further experiments, in Florida from 1970 to 1973, produced variable results owing to the occasional occurrence of unsuitably dry weather conditions, preventing conidiation of applied mycelia and reducing conidial survival (McCoy & Selhime, 1974). When the applications were successful, mite populations were maintained at low levels for 6–12 months.

Early in the 1970s, *H. thompsonii* attracted industrial attention and in 1972 Abbott Laboratories developed a commercial process for producing mycelia. Unfortunately, to prevent loss in viability, this material required cold storage, a requirement that slowed commercial development. Later, rotary drum technology was developed to produce a wettable powder containing conidia and mycelial fragments, and this formulation (Mycar) received a full registration from the Environmental Protection Agency of the USA in 1981. This formulation reduced mite populations 4 weeks after treatment to a level (9.6 per leaf) significantly lower than that on the control (26.3). In only 1 week, sporulating mycelium was produced on leaf surfaces as a result of the utilisation of nutrients in the formulation by the fungus (McCoy, 1981). However, despite the sale of several hundred kilograms of product, commercial production was stopped in 1985. The major reasons for the commercial failure of *H. thompsonii* were the requirement for cold storage, difficulties in producing infective spores, the lack of a suitable bioassay for product standardisation and inconsistent citrus rust mite control in the field (McCoy, 1986). The variable field results were most probably a result of inadequate relative humidity after application, so there is an urgent need for a formulation that allows successful use of the fungus in sub-optimal conditions.

3.12 Nomuraea rileyi: a pathogen of Lepidoptera

This fungus is primarily found parasitising Lepidoptera, though there are occasional reports from Coleoptera, e.g. *L. decemlineata* (Ignoffo, 1981). *N. rileyi* was first described over 100 years ago, but was not used for biological control until 1955 (Chamberlin & Dutky, 1958). Since then, *N.*

rileyi has been studied for control of Lepidoptera on a variety of crops, including soybeans, sweet corn and cabbage. However, research has focused mainly on the lepidopterous pest complex occurring on soybeans, including *Heliothis zea*, *Pseudoplusia includens*, *Anticarsia gemmatalis* and *Plathypena scabra*.

Growth of *N. rileyi* on agar media is unusual in that conidia germinate to form germ tubes, then subsequent growth occurs by budding, with eventual production of mycelia and conidia. Most entomogenous fungi, e.g. *B. bassiana* and *M. anisopliae*, multiply only by budding in the host, or in liquid media.

In insects, conidia germinate on the cuticle of susceptible larvae and produce slender invasion hyphae. As with most entomogenous fungi, nutrients are required for germination on the insect, which occurs optimally between 15 and 25°C (Getzin, 1961; Ignoffo *et al.*, 1976a). It is possible that infection can also occur through the gut as some 38% of *A. gemmatalis* larvae, fed with microdrops of water containing conidia, died within 8 days (Kish & Allen, 1978). Penetration of larval cuticle occurs within 24 h of conidial exposure, and larval death follows after a further 5–6 days (Ignoffo, 1981). After death, the cadaver is covered by a mycelial mat, which produces many pale-green conidia that are easily dislodged and transported by wind (Garcia & Ignoffo, 1977).

Lepidopteran species vary in their susceptibility to *N. rileyi*. Thus, *A. gemmatalis* was some 13 times less susceptible to a Mississippi isolate of *N. rileyi* than *P. scabra*. This result was consistent with field observations in Mid-Western USA soybeans that more *P. scabra* larvae were infected with *N. rileyi* than those of *A. gemmatalis* (Ignoffo, 1981). However, a Brazilian isolate had increased pathogenicity to *A. gemmatalis* and approached the virulence of the Mississippi strain in *P. scabra* (Ignoffo *et al.*, 1976c). This observation clearly demonstrates the importance of intraspecific strain selection to ensure optimal pathogenicity to a given pest.

The epizootiology of *N. rileyi* in lepidopterous insects was extensively studied by Ignoffo *et al.* (1977) who proposed a sequence of eight events:

(1) transmission of conidia surviving in the soil to plants;
(2) infection of susceptible larvae feeding on these plants;
(3) dispersal of infected larvae;
(4) larval death and conidia production;
(5) repeated infection cycles producing conidia throughout late spring and summer, with wind dispersal to initiate the late summer epizootics;
(6) death of susceptible larvae in late summer and autumn;
(7) soil contamination from conidia;
(8) overwinter survival of infectious conidia in the soil.

A model to predict the occurrence of *N. rileyi* on *A. gemmatalis* in soybean was developed by Kish & Allen (1978). In 67 field experiments conducted to test the model, Chi square values indicated a good fit of observed data and predicted values at the 5% level in 53% of the trials.

One of the most important factors governing epizootic development is relative humidity and occurrence of free water. Mortalities of *Trichoplusia ni* larvae treated with *N. rileyi* were 4, 3, 49 and 100%, at relative humidities of 42, 80, 90 and 100%, respectively (Getzin, 1961). Conidial germination can also occur in free water though an excess of heavy rain can be detrimental as it can remove conidia from both cadavers and the surrounding air, thus preventing spread by wind (Garcia & Ignoffo, 1977; Kish & Allen, 1978).

Overwinter survival of conidia is also of prime importance in the sequence of events leading to an epizootic. On soil surfaces, about 10% of activity was lost after 10–14 days, half after 40–65 days and more than 99% after 250–350 days (Ignoffo, 1981). Most conidia die during the winter, though even a mortality rate of 99.9% would still result in 6×10^6 viable conidia on 1 June for each larva that died the previous October. For average populations of larvae, this equals 5×10^4 conidia per square millimetre of soil surface (Ignoffo, 1981). A slightly higher density of 9×10^4 conidia per square millimetre of soil surface resulted in 12% mortality of *T. ni* larvae (Ignoffo *et al.*, 1977), indicating that low numbers of conidia can initiate the sequence of events leading to epizootic development of the fungus.

A number of attempts have been made to use *N. rileyi* as a mycoinsecticide. A high dose of 1.88×10^{16} conidia per hectare, applied to *Heliothis virescens* larvae feeding on tobacco, infected some larvae but 'failed to give adequate control under a considerable range of climatic conditions' (Chamberlin & Dutky, 1958). A moderate application of 3.0×10^{12} conidia per hectare killed 67% of *T. ni* larvae on cabbage, but failed to provide adequate control (Getzin, 1961). Seven weekly applications of conidia (1.4×10^{14} per hectare) gave statistically significant control of *T. ni* and reduced feeding, though cabbages still had important damage (Bell, in Ignoffo, 1981). *N. rileyi* applied to *H. zea* on soybeans at 2.5×10^{11}, 2.5×10^{12} and 2.5×10^{13} conidia per hectare reduced populations compared with those on untreated plants. However, the maximum reduction in seed-damaged pods was only 47% (Ignoffo *et al.*, 1976a). Similar results were obtained with *H. zea* on sweet corn, where application of *N. rileyi* conidia (4.0×10^{13} per hectare) killed between 61.8 and 90.2% of larvae but only reduced ear damage by a maximum 36.0% (Mohamed *et al.*, 1978). It can be concluded that the use of *N. rileyi* as a mycoinsecticide, applied frequently for control of lepidopterans on soybeans, is not a practical or economic proposition. However, it was suggested that a prophylactic application of *N. rileyi* conidia might induce an early epizootic and therefore reduce insect damage (Ignoffo *et al.*, 1975).

Evidence to support this theory was provided by Ignoffo *et al.* (1976c), who induced an epizootic in *Pl. scabra* on soybean 14 days earlier than on untreated plots by application of *N. rileyi* conidia (2.75×10^{13} per hectare) and by Sprenkel & Brooks (1975), who obtained a similar result by distributing *H. virescens* cadavers to control *H. zea, P. includens* and *Pl. scabra*. It was concluded that *N. rileyi* had potential for Lepidoptera control on soybeans if directed against young larvae, or if applied prophylactically (Ignoffo *et al.*, 1976c).

For small-scale field experiments, conidia can be produced on agar media as described by Bell (1975), who quoted yields of 6.3×10^8 conidia per square centimetre of agar surface. Thus, to produce a hectare dose of *N. rileyi* conidia, assuming the application of 2.75×10^{13} spores per hectare, would require the yield from 4.37 m^2 of agar surface, the equivalent of 192, 8.5-cm-diameter petri dishes. Clearly, such a production method is uneconomic so it is suggested that attempts be made to produce conidia on cereal grains as described for *M. anisopliae* by Aquino *et al.* (1975). Production of *N. rileyi* in liquid culture has also been attempted, with biomass yields some 1.5 times greater than those obtained on agar. Unfortunately, the blastospores produced in liquid culture were not pathogenic to larvae (Bell, 1975) which precludes their direct use as a mycoinsecticide. However, it is possible that blastospores, together with nutrients, could be applied to crops prophylactically, resulting in conidial production on the plants. These spores could then infect insects.

3.13 Use of fungi to control insect pests in the humid tropics

In natural tropical ecosystems entomogenous fungi are common and play an important role in regulating insect populations. The available records were examined by Evans (1982) who described the most frequent pathogens as *Cordyceps* spp. and reported the asexual stages (or anamorphs) as species of *Hirsutella, Hymenostilbe, Nomuraea, Paecilomyces* and *Verticillium*. Entomogenous fungi occurred frequently in a jungle in north-east Thailand, and specimens of *B. bassiana, Cordyceps* sp., *Hirsutella* sp. and *Nomuraea atypicola* were found on Diptera, Lepidoptera, jassids and spiders, respectively (A.T. Gillespie, unpublished observation).

In agricultural ecosystems fungi are less diverse, though still common on certain insects, e.g. *Hirsutella citriformis* on brown planthopper of rice (R.A. Hall unpublished observation), *Hirsutella versicolor* on mango jassids (A.T. Gillespie, unpublished observation) and *V. lecanii* (cited as *Cephalosporium lecanii*) on coffee green bug (Easwaramoorthy & Jayaraj, 1978). Occasionally entomogenous fungi kill sufficient insects to prevent them from causing economic crop damage.

The regular occurrence of fungal epizootics on tropical pests has stimulated increased research in this area, with most attention focused on the paddy rice crop. Several research groups in the UK, the Philippines and Switzerland are studying fungi for control of the brown planthopper, *Nilaparvata lugens*, and green leafhopper, *Nephotettix* spp., on rice. Field application of *M. anisopliae* and *Paecilomyces farinosus* conidia (5×10^{13} in 500 litres water per hectare) significantly reduced populations of both *N. lugens* and *Nephotettix virescens* compared with those on untreated rice plants (Gillespie *et al.*, 1986), but the degree of control obtained was inferior to the short-term control provided by chemical insecticides, and populations were reduced only slowly. However, fungi are generally specific and therefore less damaging to non-target insects than conventional insecticides, which means predators and parasites are conserved and can help to control pests. Rombach *et al.* (1986a) also demonstrated the potential of fungi to control *N. lugens*, and described high levels of infection 3 weeks after application of *M. anisopliae, M. flavoviride, H. citriformis* or *B. bassiana* at 5×10^{12} conidia per hectare. Interestingly, all fungi gave similar levels of control. In a further experiment, Rombach *et al.* (1986b) showed that *M. anisopliae, B. bassiana* and *Paecilomyces lilacinus* significantly reduced populations of the rice black bug, *Scotinophara coarctata*, for up to 9 weeks after application. Fungi have also been successfully used on coffee to control *C. viridis*. A single application of *V. lecanii* conidia (1.6×10^7 ml) killed 97.6% of treated scales 2 weeks after application, and maintained high levels of infection for up to 7 weeks (Easwaramoorthy & Jayaraj, 1978).

Generally, humidities in the tropics are much higher than in temperate regions and weather patterns are more predictable, which means fungi may be more suitable for use in the tropics. Unfortunately, most insect mycologists are based in Europe and the US and have limited or no access to the tropics. More collaborative research projects between scientists and improved training for research workers from tropical countries would go a long way towards enabling a full exploration of the potential of fungi to control tropical pests.

3.14 Conclusions

Having examined the available data on the practical use of fungi for the control of agricultural pests, it must be concluded that limited progress has been made during the last 100 years. However, knowledge is now accumulating rapidly. The reliability of fungi for pest control is still very dependent on the prevailing environmental conditions after application of spores. Since fungi require high moisture levels for spore germination and subsequent mycelial growth, this is not surprising.

There is an urgent need for research to develop moisture-retaining formulations which allow fungal growth at sub-optimal relative humidities. Published data is uncommon: the involvement of commercial companies in mycoinsecticide development unavoidably means certain information remains industrial property and so is not generally available. However, there have been recent developments in the formulation of biological agents to control plant disease. Philipp & Hellstern (1986) reported that an emulsion of liquid paraffin allowed the fungus *Ampelomyces quisqualis* to parasitise powdery mildew mycelium at relative humidities as low as 80%. Similar formulations should be studied for use with entomogenous fungi. However, even if effective formulations can be devised, they can assist only initial infection and any subsequent spread of disease will depend on favourable conditions occurring naturally.

Particularly until moisture-retaining formulations are available, careful consideration needs to be given to the pest species targeted as likely candidates for control with entomogenous fungi: there is no point in developing a mycoinsecticide for use in situations where relative humidities are unfavourable for fungal growth. This approach has been adopted by some insect mycologists and, increasingly, research has been focused on insects occurring in humid microclimates, e.g. soil and rice pests, stem borers and spittle bugs (Cercopidae). Soil is a particularly suitable environment for exploitation of fungi, as moisture levels are usually sufficient for fungal germination and growth occurs even at the permanent wilting point for mesophytic higher plants (pF 4.2; equivalent to about 98% r.h.; Griffin, 1963). Although application of fungal spores to soil removes the important constraint of limiting relative humidities, spores have to germinate in competition with a plethora of other soil microorganisms. However, the entomogenous fungi's specific adaptation to parasitisation of insects probably confers an important ecological advantage.

Other important physical factors which limit the use of fungi are sunlight and temperature. Sunlight can reduce spore survival. Ignoffo (1976b) described the half-life of *N. rileyi* conidia on soybean leaves as 2–3 days, and Zimmermann (1982) reported a half-life of 2–3 h for *M. anisopliae* conidia. Obviously, conidia in soil will receive some protection from sunlight as will those applied to the undersurface of leaves. Electrostatic sprayers provide good coverage of both leaf surfaces and might be useful for field application of fungi. Recent glasshouse experiments, examining *V. lecanii* for control of *Aphis gossypii* on chrysanthemums, demonstrated that ultra-low-volume application of *V. lecanii* with an electrostatic sprayer at 10 litres per hectare provided control superior to that obtained with conventional high-volume application at 1000 litres per hectare. (P. Sopp & A.T. Gillespie, unpublished observation). Many isolates of *B. bassiana* and *M. anisopliae* have optimum temperatures for conidial germination and mycelial growth of from 25 to 28°C, and certain isolates of *M.*

anisopliae grow only slowly at 20°C (Gillespie, 1984). Whereas temperatures of 25°C and above occur commonly in the tropics, such conditions occur only rarely in temperate areas. It is therefore important to select fungal isolates able to grow rapidly at the temperatures prevailing after spore application. Rapid germination allows fungal establishment during limited periods of high relative humidity, and reduces the likelihood of moulting insects shedding the inoculum on the cast skins, as shown by Fargues & Vey (1974).

Strain selection is of great importance when choosing a fungal isolate for control of a given pest. Important considerations are pathogenicity to the target insect, sporulation on dead insects, disease spread, ease of production in suitable media and survival of infective propagules. Fungal strains can vary widely in all these respects, and it is perhaps best to regard each strain as an individual entity and not generalise within a species. Furthermore, a wild-type isolate comprises a number of different genotypes which can vary markedly in characteristics such as pathogenicity (Samsinakova & Kalalova, 1983; J. Jimenez & A.T. Gillespie, unpublished observation). Thus it is recommended that selection be made of both strains and single-spore isolates.

Once the best available strain has been selected, attention must be given to production and formulation. Any proposal to produce conidia on agar is likely to be hopelessly uneconomic. The production method of choice is by liquid fermentation, utilising conventional stirred tank reactors. Some fungi, e.g. *V. lecanii*, produce many blastospores in liquid culture and these can be harvested by centrifugation, dried and formulated. However, these spores are relatively short-lived, being adapted for fungal dissemination within an insect rather than survival, and require cool storage to maintain viability. Other fungi, e.g. *M. anisopliae*, produce mainly mycelia in liquid culture, and maximum spore yields are about 10^8 per millilitre (Adamek, 1963; Gillespie, 1984). The mycelium and spores can be formulated and dried using the technique of McCabe & Soper (1985). However, storage problems and delays in pest control provide significant disadvantages for this method. *M. anisopliae* can be produced economically on cereal grains in sterilisable polyethylene bags, or could probably be produced using rotary drum technology as developed for *H. thompsonii*. After sporulation, the cereal grains can be washed to harvest the spores, which are dried, ground and formulated to produce a wettable powder.

Certain isolates of *B. bassiana* conidiate in liquid culture and produce maximum yields of 5×10^8 spores per millilitre (Thomas *et al.*, 1986; U. Hilber & A.T. Gillespie, unpublished observation). These conidia were pathogenic to test insects and although yields were some 10 times lower than those obtained on barley grain the results suggest that further research could be worthwhile, though storage of liquid-produced conidia might present problems. Conidia of *B. bassiana* produced on agar survived well at moderate temperatures and one formulation maintained over 90% viability for at least 3 months, at 20°C (J. Jimenez & A.T. Gillespie, unpublished observation). To date, only limited research has been conducted on developing fungal formulations which store for long periods. Insect mycologists could perhaps learn from food microbiologists, who have successfully developed stable formulations of yeasts.

The emerging techniques of molecular biology present exciting possibilities for future study. However, it must be remembered that these new techniques can be utilised only when a fuller understanding of factors governing pathogenicity of fungi to insects is obtained. Progress is being made, particularly with respect to enzymic degradation of insect cuticle, where the key enzyme is probably an endoprotease (St. Leger *et al.*, 1986a). Possibly the insertion of multiple copies of the gene coding for this enzyme would improve the efficiency of cuticular penetration by fungi. Other possible strategies are to insert toxin-coding genes into fungi to increase the speed of kill and to bring together many desirable features into single strains.

However, none of these strategies will realise its full potential for pest control unless formulations are obtained to reduce the reliance of fungi on high levels of relative humidity. If this goal can be achieved, the use of fungi for control of agricultural pests may become common and agriculture will benefit from the advantages of prolonged pest control, a relatively narrow spectrum of activity, a reduced risk of insect resistance and a high degree of safety to non-target organisms.

References

Adamek, L. (1963). Submerged cultivation of the fungus, *Metarhizium anisopliae* (Metsch.). *Folia Microbiologie (Praha)*, **10**, 255–7.
Aquino, de M.L., Cavalanti, V.A., Sena, R.C. & Queiroz, G.F. (1975). Nova technologia de multipliçacào do fungo *Metarhizium anisopliae*. *Boletim Técnico da Comissào Executiva de Defesa Fitossanitaria da Lavoura Canavieira de Pernambuco*, **4**, 1–31.
Bedford, G.O. (1980). Biology, ecology, and control of palm rhinoceros beetles. *Annual Review of Entomology*, **25**, 309–39.
Bell, J.V. (1975). Production and pathogenicity of the fungus *Spicaria rileyi* from solid and liquid media. *Journal of Invertebrate Pathology*, **26**, 129–30.
Ben Ze'ev, I. & Kenneth, R.G. (1982a). Features-criteria of taxonomic value in the Entomophthorales: I A revision of the Batkoan system. *Mycotaxon*, **14**, 393–455.
Ben Ze'ev, I. & Kenneth, R.G. (1982b). Features-criteria of taxonomic value in the Entomophthorales: II A revision of the genus *Erynia* Nowakowski 1881 (= *Zoophthora* Batko 1964). *Mycotaxon*, **14**, 456–75.
Berger, E.W. (1909). Whitefly studies in 1908. *Bulletin of the Florida Agricultural Experiment Station*, **97**, 42–59.

Blachère, H., Calves, J., Ferron, P., Corrieu, G. & Peringer, P. (1973). Etude de la formulation et de la conservation d'une préparation entomopathogène à base de blastospores de *Beauveria tenella* (Delacr. Siemasko). *Annales du Zoologie et Ecologie Animale*, **11**, 247–57.

Carl, K.P. (1975). An *Entomophthora* sp. [Entomophthorales: Entomophthoraceae] pathogenic to *Thrips* spp, [Thysan.: Thripidae] and its potential as a biological control agent in glasshouses. *Entomophaga*, **20**, 381–8.

Chamberlin, F.S. & Dutky, S.R. (1958). Tests of pathogens for the control of tobacco insects. *Journal of Economic Entomology*, **51**, 560.

Cinovskis, J., Cudare, Z., Jegina, K., Petrova, V. & Strazdina, A. (1974). The use of the fungus *Entomophthora thaxteriana* Petch. against apple tree aphids. *Izvestiya Akademii Nauk Latvilskoi SSR*, **7**, 33–6.

Coles, R. & Pinnock, D.E. (1984). Current status of the production and use of *Metarhizium anisopliae* for control of *Aphodius tasmaniae* in South Australia. *Proceedings of the IVth Australian Applied Entomological Research Conference*, eds P. Bailey & D. Swincer, pp. 357–61, South Australia Government Printer, Adelaide.

Easwaramoorthy, S. & Jayaraj, S. (1978). Effectiveness of the white halo fungus, *Cephalosporium lecanii*, against field populations of coffee green bug, *Coccus viridis*. *Journal of Invertebrate pathology*, **32**, 88–96.

Evans, H.C. (1982). Entomogenous fungi in tropical forest ecosystems: an appraisal. *Ecological Entomology*, **7**, 47–60.

Evans, H.C. & Samson, R.A. (1986). The genus *Verticillium*: taxonomic problems in species with invertebrate hosts. In: *Fundamental and Applied Aspects of Invertebrate Pathology*, eds R.A. Samson, J.M. Vlak & D. Peters. Foundation of the IVth International Colloquium of Invertebrate Pathology, Wageningen, The Netherlands, pp. 186–189.

Fargues, J. & Robert, P.H., (1985). Persistence des conidiospores des hyphomycetes entomopathogenes *Beauveria bassiana* (Bals.) Vuill., *Metarhizium anisopliae* (Metsch.) Sor., *Nomuraea rileyi* (F.) Samson et *Paecilomyces fumosoroseus* Wize dans le sol, en conditions contrôlées. *Agronomie*, **5**, 73–80.

Fargues, J. & Vey, A. (1974). Modalités d'infection des larves de *Leptinotarsa decemlineata* par *Beauveria bassiana* au cours de la mue. *Entomophaga*, **19**, 311–23.

Fargues, J., Reisinger, O., Robert, P.H. & Aubart, C. (1983). Biodegradation of entomopathogenic hyphomycetes: influence of clay coating on *Beauveria bassiana* blastospore survival in soil. *Journal of Invertebrate Pathology*, **41**, 131–42.

Ferron, P. (1978). Etiologie et épidémiologie des muscardines. Thèse Doctorat d'Etat, Université Pierre et Marie Curie, Paris, 294 pp.

Ferron, P. (1981). Pest control by the fungi *Beauveria* and *Metarhizium*. In: *Microbial Control of Pests and Plant Diseases 1970–1980*, ed. H.D. Burges, pp. 465–82. Academic Press, London.

Franz, J.M. & Krieg, A. (1980). Mikrobiologische Schädlingsbekämpfung in China. *Ein Reisebert Forum Mikrobiologie*, **3**, 173–6.

Garcia, C. & Ignoffo, C.M. (1977). Dislodgement of conidia of *Nomuraea rileyi* from cadavers of cabbage looper, *Trichoplusia ni*. *Journal of Invertebrate Pathology*, **30**, 114–6.

Getzin, L.W. (1961). *Spicaria rileyi* (Farlow) Charles, an entomogenous fungus of *Trichoplusia ni* (Hübner). *Journal of Insect Pathology*, **3**, 2–10.

Gillespie, A.T. (1984). The potential of entomogenous fungi to control glasshouse pests and brown planthopper of rice. PhD Thesis University of Southampton, 154 pp.

Gillespie, A.T. & Crawford, E. (1986). Effect of water activity on conidial germination and mycelial growth of *Beauveria bassiana*, *Metarhizium anisopliae*, *Paecilomyces* spp. and *Verticillium lecanii*. In: *Fundamental and Applied Aspects of Invertebrate pathology*, eds R.A. Samson, J.M. Vlak & D. Peters, p. 254. Foundation of the IVth International Colloquium of Invertebrate Pathology, Wageningen, The Netherlands.

Gillespie, A.T., Collins, M.D. & Atienza, A. (1986). Control of *Nilaparvata lugens* with entomogenous fungi. In: *Fundamental and Applied Aspects of Invertebrate Pathology*, eds R.A. Samson, J.M. Vlak & D. Peters, p. 244. Foundation of the IVth International Colloquium of Invertebrate Pathology, Wageningen, The Netherlands.

Goral, V.M. (1979). Effect of cultivation conditions on the entomopathogenic properties of muscardine fungi. In: *Proceedings, First Joint US/USSR Conference*, Kiev, pp. 217–24.

Griffin, D.M. (1963). Soil moisture and the ecology of soil fungi. *Biological Review*, **38**, 141–66.

Hänel, H. (1982). Propagation of *Metarhizium anisopliae* infection in termite colonies in the laboratory and in the field. *Proceedings of the IIIrd International Colloquium of Invertebrate Pathology*, Papers, 107.

Hughes, J.C. & Gillespie, A.T. (1985). Germination and penetration of the aphid *Macrosiphoniella sanborni* by two strains of *Verticillium lecanii*. *Programme and Abstracts, XVIIIth Annual Meeting of the Society for Invertebrate Pathology*, Sault Ste Marie, p. 28.

Humber, R.A. (1981). An alternative view of certain taxonomic criteria used in the Entomophthorales (Zygomycetes). *Mycotaxon*, **13**, 191–240.

Hurpin, B. & Robert, P.H. (1972). Comparison of the activity of certain pathogens of the cockchafer, *Melolontha melolontha* in plots of natural meadowland. *Journal of Invertebrate Pathology*, **19**, 291–8.

Hussey, N.W. & Tinsley, (1981). Impressions of insect pathology in the People's Republic of China. In: *Microbial Control of Pests and Plant Diseases 1970–1980*, ed. H.D. Burges, pp. 785–95. Academic Press, London.

Huxham, I.M., Lackie, A.M. & McCorkindale, N.J. (1986). An *in vitro* assay to investigate activation and suppression by a pathogenic fungus of prophenol oxidase by insect haemocytes. In: *Fundamental and Applied Aspects of Invertebrate Pathology*, eds R.A. Samson, J.M. Vlak & D. Peters, p. 463. Foundation of the IVth International Colloquium of Invertebrate Pathology, Wageningen, The Netherlands.

Ignoffo, C.M. (1981). The fungus *Nomuraea rileyi* as a microbial insecticide. In: *Microbial Control of Pests and Plant Diseases 1970–1980*, pp. 513–38. Academic Press, London.

Ignoffo, C.M., Garcia, C. & Hostetter, D.L. (1976a). Effects of temperature on growth and sporulation of the entomopathogenic fungus *Nomuraea rileyi*. *Environmental Entomology*, **5**, 935–6.

Ignoffo, C.M., Garcia, C., Hostetter, D.L. & Pinnell, R.E. (1977). Laboratory studies of the entomopathogenic fungus *Nomuraea rileyi*: soil-borne contamination of soybean seedlings and dispersal of diseased larvae of *Trichoplusia ni*. *Journal of Invertebrate Pathology*, **29**, 147–52.

Ignoffo, C.M., Hostetter, D.L., Biever, K.D., Garcia, C., Thomas, G.D., Dickerson, W.A. & Pinnell, R.E. (1978). Evaluation of an entomopathogenic bacterium, fungus and virus for control of *Heliothis zea* on soybeans. *Journal of Economic Entomology*, **71**, 165–8.

Ignoffo, C.M., Marston, N.L., Hostetter, D.L. & Puttler, B. (1976b). Natural and induced epizootics of *Nomuraea rileyi* in soybean caterpillars. *Journal of Invertebrate Pathology*, **27**, 191–8.

Ignoffo, C.M., Puttler, B., Hostetter, D.L. & Dickerson, W.A. (1976c). Susceptibility of the cabbage looper, *Trichoplusia ni*, and the velvetbean caterpillar, *Anticarsia gemmatalis*, to several isolates of the entomopathogenic fungus *Nomuraea rileyi*. *Journal of Invertebrate Pathology*, **28**, 259–62.

Ignoffo, C.M., Puttler, B., Marston, N.L., Hostetter, D.L. & Dickerson, W.A. (1975). Seasonal incidence of the entomopathogenic fungus *Spicaria rileyi* associated with noctuid pests of soybeans. *Journal of Invertebrate Pathology*, **25**, 135–7.

Keller, S. (1978). Infektionsversuche mit dem Pilz *Beauveria tenella* an adulten Maikäfern (*Melolontha melolontha* L.). *Mitteilungen der Schweizerischen Entomologischen Gesellschaft*, **51**, 13–19.

Keller, S. (1979). Ergebnisse eines Versuches zur mikrobiologischen Bekämpfung des Maikäfers (*Melolontha melolontha* L.) mit dem Pilz *Beauveria tenella*. *Mitteilungen der Schweizerischen Entomologischen Gesellschaft*, **52**, 35–44.

Keller, S. (1981). Field trials of *Conidiobolus obscurus* against pea aphids in Switzerland. In: *Euraphid 1980. Aphid Forecasting and Pathogens and a Handbook for Aphid Identification*, ed. L.R. Taylor, p. 42. Rothamsted Experimental Station, UK.

Keller, S. (1983). Die mikrobiologische Bekämpfung des Maikäfers (*Melolontha melolontha* L.) mit dem Pilz *Beauveria brongniartii*. *Mitteilungen der Schweizerischen Landwirtschaft*, **31**, 61–4.

Keller, S. (1986). Control of may beetle grubs (*Melolontha melolontha* L.) with the fungus *Beauveria brongniartii* (Sacc.) Petch. In: *Fundamental and Applied Aspects of Invertebrate Pathology*, eds R.A. Samson, J.M. Vlak & D. Peters, pp. 525–8. Foundation of the IVth International Colloquium of Invertebrate Pathology, Wageningen, The Netherlands.

Kelsey, J.M. (1965). *Entomophthora sphaerosperma* (Fres.) and *Plutella maculipennis* (Curtis) control. *New Zealand Entomologist*, **3**, 47–9.

Kish, L.P. & Allen, G.E. (1978). The biology and ecology of *Nomuraea rileyi* and a program for predicting its incidence on *Anticarsia gemmatalis* in soybean. *Florida Agricultural Experimental Station Bulletin*, **795**, 48 pp.

Krassilstschik, J. (1888). La production industrielle des parasites vegetaux pour la destruction des insectes nuisible. *Bulletin Science France, Belgique*, **19**, 461–72.

Kravtsov, A.A. (1982). Biological preparations. *Zashchita Rastenii*, **6**, 58–61.

Latgé, J.P. (1982). Production of Entomophthorales. *Proceedings of the IIIrd International Colloquium of Invertebrate Pathology*, Papers, 164–9.

Latgé, J.P. (1986). The Entomophthorales after the resting spore production stage. In: *Fundamental and Applied Aspects of Invertebrate Pathology*, eds R.A. Samson, J.M. Vlak & D. Peters, pp. 651–2. Foundation of the Fourth International Colloquium of Invertebrate Pathology, Wageningen, The Netherlands.

Latgé, J.-P., Remaudière, G., Soper, R.S. Madore, C.D. & Diaquin, M. (1978). Growth and sporulation of *Entomophthora virulenta* on semidefined media in liquid culture. *Journal of Invertebrate Pathology*, **31**, 225–33.

Latgé, J.P., Soper, R.S. & Madore, C.D. (1977). Media suitable for industrial production of *Entomophthora virulenta* zygospores. *Biotechnology and Bioengineering*, **19**, 1269–84.

Lingg, A.J. & Donaldson, M.D. (1981). Biotic and abiotic factors affecting stability of *Beauveria bassiana* in soil. *Journal of Invertebrate Pathology*, **38**, 191–200.

Lipa, J.J. (1985). Progress in biological control of the Colorado beetle (*Leptino-*

tarsa decemlineata) in Eastern Europe. *Bulletin E.P.P.O.*, **15**, 207–11.

McCabe, D. Soper, R.S. (1985). Preparation of an entomopathogenic fungal insect control agent. *United States Patent*, 4,530,834, July 23, 1985, pp. 1–4.

McCoy, C.W. (1981). Pest control by the fungus *Hirsutella thompsonii*. In: *Microbial Control of Pests and Plant Diseases 1970–1980*, ed. H.D. Burges, pp. 499–512. Academic Press, London.

McCoy, C.W. (1986). Factors governing the efficacy of *Hirsutella thompsonii* in the field. In: *Fundamental and Applied Aspects of Invertebrate Pathology*, eds R.A. Samson, J.M. Vlak & D. Peters, pp. 171–4. Foundation of the IVth International Colloquium of Invertebrate Pathology, Wageningen, the Netherlands.

McCoy, C.W. & Selhime, A.G. (1974). The fungus pathogen, *Hirsutella thompsonii* and its potential use for control of the citrus rust mite in Florida. *Proceedings of the International Citrus Congress*, **2**, 521–7.

McCoy, C.W., Selhime, A.G., Kanavel, R.F. & Hill, A.J. (1971). Suppression of citrus rust mite populations with application of fragmented mycelia of *Hirsutella thompsonii*. *Journal of Invertebrate Pathology*, **17**, 270–6.

Metchnikoff, E. (1879). Maladies des hannetons duble. *Zapiski imperatorskogo obshcestaa sel'skago Khozyaistra yuzhnoi rossii*, 17–50.

Milner, R.J. & Lutton, G.G. (1986). Dependence of *Verticillium lecanii* (Fungi: Hyphomycetes) on high humidities for infection and sporulation using *Myzus persicae* (Homoptera: Aphididae) as host. *Environmental Entomology*, **15**, 380–2.

Milner, R.J., Soper, R.S. & Lutton, G.G. (1982). Field release of an Israeli strain of the fungus *Zoophthora radicans* (Brefeld) Batko for biological control of *Therioaphis trifolii* (Monell) *f. maculata*. *Journal of the Australian Entomological Society*, **21**, 113–18.

Mohamed, K.A., Bell, J.V. & Sikorowski, P. (1978). Field cage tests with *Nomuraea rileyi* against corn earworm larvae on sweet corn. *Journal of Economic Entomology*, **71**, 102–4.

Perry, D.F., Latteur, G. & Wilding, N. (1982). The environmental persistence of propagules of the Entomophthorales. *Proceedings of the IIIrd International Colloquium of Invertebrate Pathology*, Papers, 164–9.

Philipp, W.-D. & Hellstern, A. (1986). Biologische Mehltanbekampfung mit *Ampelomyces quisqualis* bei reduzierter Luftfeuchtigkeit. *Zeitschrift für Pflanzenkrankheiten und Pflanzenschutz*, **93**, 384–91.

Poprawski, T.J., Robert, P.-H., Majchrowicz, I. & Boivin, G. (1985). Susceptibility of *Delia antiqua* (Diptera: Anthomyiidae) to eleven isolates of entomopathogenic hyphomycetes. *Environmental Entomology*, **14**, 557–61.

Remaudière, G. (1981). Provisional field trials with *Conidiobolus obscurus* and proposed trials for 1981. In: *Euraphid 1980. Aphid Forecasting and pathogens and a Handbook for Aphid Identification*, ed. L.R. Taylor, pp. 41–2. Rothamsted Experimental Station, UK.

Remaudière, G. & Keller, S. (1980). Révision systématique des genres d'Entomophthoraceae à potentialité entomopathogène. *Mycotaxon*, **11**, 323–38.

Remaudière, G., Latgé, J.P. & Michel, M.F. (1981). Ecologie comparée des Entomophthoracées pathogènes de pucerons en France littorale et continentale. *Entomophaga*, **26**, 157–78.

Riba, G. (1984). Application en éssais parcellaires de plein champ d'un mutant artificiel du champignon entomopathogène *Beauveria bassiana* (Hyphomycete) contre la pyrale du mais, *Ostrinia nubilalis* [Lep.: Pyralidae]. *Entomophaga*, **29**, 41–8.

Roberts, D.W. (1981). Toxins of entomopathogenic fungi. In: *Microbial Control of Pests and Plant Diseases 1970–1980*, ed. H.D. Burges, pp. 441–64. Academic Press, London.

Roberts, D.W. & Wraight, S.P. (1986). Current status on the use of insect pathogens as biocontrol agents in agriculture: fungi. In: *Fundamental and Applied Aspects of Invertebrate pathology*, eds R.A. Samson, J.M. Vlak & D. Peters, pp. 510–13. Foundation of the IVth International Colloquium of Invertebrate Pathology, Wageningen, The Netherlands.

Rolfs, P.H. & Fawcett, H.S. (1908). Fungus diseases of scale insects and whitefly. *Bulletin of the Florida Agricultural Experiment Station*, **94**, 1–17.

Rombach, M.C., Aguda, R.M., Shepard, B.M. & Roberts, D.W. (1986a). Infection of rice brown planthopper, *Nilaparvata lugens* (Homoptera: Delphacidae), by field application of entomopathogenic hyphomycetes (Deuteromycotina). *Environmental Entomology*, **15**, 1070–3.

Rombach, M.C., Aguda, R.M., Shepard, B.M. & Roberts, D.W. (1986b). Entomopathogenic fungi (Deuteromycotina) in the control of the black bug of rice, *Scotinophara coarctata* (Hemiptera, Pentatomidae). *Journal of Invertebrate Pathology*, **48**, 174–9.

Samsinakova, A. & Kalalova, S. (1983). The influence of single spore isolate and repeated subculturing on the pathogenicity of conidia of the entomophagous fungus *Beauveria bassiana*. *Journal of Invertebrate Pathology*, **42**, 156–61.

Shands, W.A., Hall, I.M. & Simpson, G.W. (1962). Entomophthoraceous fungi attacking the potato aphid in Northeastern Maine in 1960. *Journal of Economic Entomology*, **55**, 174–9.

Sikura, A.I. & Sikura, L.V. (1983). Use of biopreparations. *Zashchita Rastenii*, **5**, 38–9.

Smith, R.J., Pekrul, S. & Grula, E.A. (1981). Requirement for sequential enzymatic activities for penetration of the integument of the corn earworm (*Heliothis zea*). *Journal of Invertebrate pathology*, **38**, 335–44.

Sprenkel, R.K. & Brooks, W.M. (1975). Artificial dissemination and epizootic initiation of *Nomuraea rileyi*, an entomogenous fungus of lepidopterous pests of soybeans. *Journal of Economic Entomology*, **68**, 847–50.

St. Leger, R.J., Charnley, A.K. & Cooper, R.M. (1986a). Cuticle-degrading enzymes of entomopathogenic fungi: synthesis in culture on cuticle. *Journal of Invertebrate Pathology*, **48**, 85–95.

St. Leger, R.J., Cooper, R.M. & Charnley, A.K. (1986b). Cuticle-degrading enzymes of entomopathogenic fungi: cuticle degradation *in vitro* by enzymes from entomopathogens. *Journal of Invertebrate Pathology*, **47**, 167–77.

St. Leger, R.J., Cooper, R.M. & Charnley, A.K. (1987). Production of cuticle-degrading enzymes by the entomopathogen *Metarhizium anisopliae* during infection of cuticles from *Calliphora vomitoria* and *Manduca sexta*. Journal of General Microbiology, **133**, 1371–82.

Thomas, K.C., Khachatourians, G.G. & Ingledew, W.M. (1987). Production and properties of *Beauveria bassiana* conidia cultivated in submerged culture. *Canadian Journal of Microbiology*, **33**, 12–20.

Tulloch, M. (1976). The genus *Metarhizium*. *Transactions of the British Mycological Society*, **66**, 407–11.

Van Winkelhoff, A.J. & McCoy, C.W. (1984). Conidiation of *Hirsutella thompsonii* var. *synnematosa* in submerged culture. *Journal of Invertebrate Pathology*, **43**, 59–68.

Watt, B.A. & Lebrun, R.A. (1984). Soil effects of *Beauveria bassiana* on pupal populations of the Colorada potato beetle (Coleoptera: Chrysomelidae). *Environmental Entomology*, **13**, 15–18.

Wilding, N. (1973). The survival of *Entomophthora* spp. in mummified aphids at different temperatures and humidities. *Journal of Invertebrate Pathology*, **21**, 309–11.

Wilding, N. (1981). Pest control by Entomophthorales. In: *Microbial Control of Pests and Plant Diseases 1970–1980*, ed. H.D. Burges, pp. 539–54. Academic Press, London.

Wilding, N. (1982). Entomophthorales: field use and effectiveness. *Proceedings of the IIIrd International Colloquium of Invertebrate Pathology*, Papers 170–5.

Zacharuk, R.Y. (1970a). Fine structure of *Metarhizium anisopliae* infecting three species of larval Elateridae (Coleoptera) I. Dormant and germinating conidia. *Journal of Invertebrate Pathology*, **15**, 63–80.

Zacharuk, R.Y. (1970b). Fine structure of *Metarhizium anisopliae* infecting three species of larval Elateridae (Coleoptera) II. Conidial germ tubes and appressoria. *Journal of Invertebrate Pathology*, **15**, 81–91.

Zacharuk, R.Y. (1970c). Fine structure of the fungus *Metarhizium anisopliae* infecting three species of larval Elateridae (Coleoptera) III. Penetration of the host integument. *Journal of Invertebrate Pathology*, **15**, 372–96.

Zimmermann, G. (1982). Effect of high temperatures and artificial sunlight on the viability of conidia of *Metarhizium anisopliae*. *Journal of Invertebrate Pathology*, **40**, 36–40.

Mass production of entomogenous fungi for biological control of insects

4.1 Introduction

More than 150 years have passed since Augustino Bassi proposed micro-organisms as biological control agents of insect pests. Thirty-seven years later, in 1873, LeConte recommended the deliberate study of insect diseases to control noxious insects. In the same year, Eli Metchnikoff proposed the use of the fungus *Metarhizium anisopliae* for control of sugar beet curculio and succeeded in releasing laboratory-grown fungus against the pest.

Since then, a number of fungi have been discovered to be prime candidate biocontrol agents (Table 4.1). Some have been developed commercially (*Beauveria* in the USSR and Czechoslovakia; *Metarhizium* in Brazil; *Hirsutella* in the US), and other species still require varying degrees of research and development.

The scientific validity of using a fungal pathogen against an insect does not equal technical practicality: there are problems of efficacy under unfavourable conditions, stability of formulations, compatibility with current application methods, and mass production. Discussions of such obstacles have been presented by Lisansky & Hall (1983) and Jaronski (1986).

Table 4.1. Fungi under consideration as biological control agents

Fungus	In vitro culture	Pilot scale	Commercial production
Oomycetes			
Lagenidium giganteum	———————→		
Chytridiomycetes			
Coelomomyces spp.	—		
Zygomycetes			
Entomophthorales[a]	———————→		
Aschersonia aleyrodis	→		
Deuteromycetes			
Beauveria bassiana	———————————————→		
B. brogniartii	———————→		
Metarhizium anisopliae	———————————————→		
Paecilomyces farinosus	——————→		
P. fumoso-roseus	——————→		
Verticillium lecanii	———————————————→		
Nomuraea rileyi	——————→		
Hirsutella thompsonii	———————————————→		
Tolypocladium cylindrosporum	→		
Culicinomyces clavosporus	→		

[a] Only a few species have been cultured *in vitro*.

The fundamental consideration in potential commercialisation of any of these fungi is whether an efficacious product can be developed and produced cheaply enough to compete with existing controls, especially chemicals, to return an acceptable profit on the research and development investment. Low-cost mass production of an entomogenous fungus is only one of a number of technical constraints. Yet mass production is basic to commercialisation, and so it is the subject of our discussion here.

4.2 Methods of mycoinsecticide production

Entomogenous fungi fall into four general groups: the Oomycetes, the Zygomycetes (Entomophthorales), the Chytridiomycetes (*Coelomomyces*) and the Deuteromycetes. The first three groups tend to be specialised pathogens with narrow host ranges and fastidious nutritional needs. The Coelomomycetaceae in addition require an intermediate host for survival in nature.

The Deuteromycetes on the other hand, are much broader in their host spectra and are able to grow and sporulate on more generalised media. A number of these fungi may really be soil saprophytes that have been selected for the exploitation of a unique substrate — the soil-dwelling insect — not available to most other soil microorganisms. The generally non-fastidious nature of these insect-pathogenic Deuteromycetes has made them the most amenable to mass production.

What follows is a survey of past methods used to mass produce the important entomogenous fungi in order to give the reader a better perspective on the problems encountered.

4.2.1 *Oomycetes*

The only Oomycete currently being examined for large-scale biocontrol is *Lagenidium giganteum*, a pathogen of mosquito larvae. For a review of this and related Oomycetes see Jaronski (1982).

Two approaches to mass production have been followed: production of mycelium competent to produce the infective zoospores when immersed in water, and oospores, which are not infective in themselves but can withstand long periods of storage in a dry condition and give rise to zoospores when rewetted.

L. giganteum has distinct sterol requirements, namely cholesterol, ergosterol and campesterol (Domnas *et al.*, 1977). After repeated culture using sterol-deficient media, such as Emerson's peptone–yeast–glucose agar (PYG), the fungus loses its ability to produce zoospores. Oospore production is even more sensitive to absence of sterols (Kerwin & Washino, 1983).

Jaronski & Axtell (1984) approached mass production with a two-phase system. Mycelium is grown in shake flasks of PYG, then used to inoculate large shake-flasks of aqueous sunflower seed or cotton seed extract. The mycelium is harvested after a week, broken up by gentle homogenisation, and plated on agar containing the seed extract. *Lagenidium* colonises the agar substrate and becomes competent to produce the infective zoospores after three days. Zoosporogenesis is initiated when these agar cultures are immersed in water. The authors found that two standard petri dishes of *Lagenidium* culture (*c.* 160 cm^2 agar surface) were sufficient to treat 1 m^2 of mosquito habitat. More recently Su *et al.* (1986) reported that the agar cultures of *Lagenidium* could be dried to 10% moisture and stored for up to three months at room temperature with little decrease in zoosporogenic ability. These methods, however, are only practical for small field tests or local production under primitive conditions.

Oospore production has been addressed by Kerwin & Washino (1986). Calcium, unsaturated fatty acids, and sterols are the key components of their liquid medium. The latter two groups may be supplied as vegetable oil–lecithin mixtures. Current yields have been of the order of 5×10^7 oospores per litre of medium in shake flasks or small fermenters.

Dry oospores will remain viable for many months, then germinate and produce infective zoospores (15–20 zoospores per oospore) when wetted. These attributes give oospores considerable commercial potential. Oosporogenesis, however, is easily affected by nutritional and physical variations during fermentation (Latgé *et al.*, 1986).

Both methods, while sufficient to provide material for small field tests, are clearly insufficient for commercial mass production. Additional effort must be expended to scale either of these methods to a level acceptable to industry.

4.2.2 *Zygomycetes*

4.2.2.1 *Entomophthorales.* The Entomophthorales are a large and very diverse group of host-specific insect pathogens. Variations in growth requirements, host specificity, biology, etc. are so great that generalisations are ill-advised. The biology and importance of this group are summarised by Wilding (1981) and Soper (1985).

The structures important for biocontrol are the aerial conidium and the resting spore. Resting spores are the most important infectious particles, but the phenology and biology of this stage are poorly understood.

Many species have not yet been cultured *in vitro*. Successful isolations and subculture have generally required either an egg-yolk agar or protoplast culture in modified insect cell culture media.

A few potentially important entomophthorans, however, have been cultured on a pilot scale (Latgé *et al.*, 1978; Soper, 1978; Latgé & Perry, 1980). The techniques that have been developed generally involve submerged fermenter production of resting spores, which are then spray-dried (Latgé *et al.*, 1977; Soper, 1978).

Dextrose-yeast extract and soybean flour–yeast extract gave optimal *Entomophthora virulenta* resting spore production ($2.0–4.5 \times 10^9$ per litre) in 14-litre fermenters (Latgé *et al.*, 1977). The product was a milled wettabale powder containing resting spores. *Conidiobolus obscurus* resting spores were optimally produced in unrefined corn oil or glucose and corn steep liquor with yields of 4×10^9 spores –1 per litre after 60 (Latgé & Perry, 1980). The subject entomophthorans are admittedly easy to grow and readily sporulate on solid and in liquid media. This is not the case with most other entomophthorans. Resting spores of many species require several months of vernalisation at 4–7°C (Latgé & Perry, 1980). Only after suitable conditioning are the spores competent to produce infectious conidia under suitable environmental conditions.

A two-phase system, bypassing the resting spore stage, has recently been patented (McCabe & Soper, 1985). In this protocol, mycelium is produced in submerged fermentation, dried with a protective agent (generally a sugar), then ground into a powder. When rewetted, the mycelium quickly sporulates under suitable environmental conditions to produce infectious spores. Problems still remain: the method is restricted to those few entomophthorans amenable to culture on artificial media; the mycelial product has to be stored at or below 4°C to maintain its viability, and the conidia produced by reconstituted mycelium are also short-lived and fragile. These many problems have led at least one worker to conclude that environmental manipulations to conserve and encourage natural epizootics are preferable to inundative release of a mycoinsecticide (Latgé, 1986).

4.2.2.2 *Aschersonia aleyrodis*.

This is a specific pathogen of whiteflies (Homoptera: Aleyrodidae) that has attracted the attention of biological researchers since the late 1960s. It differs from the other entomopathogenic fungi in that conidia are formed within pycnidia rather than on aerial structures.

Current research indicates that application of about 10^8 conidia per mature cucumber plant are necessary to achieve control in glasshouses (Ramakers and Samson, 1984). Conidia can be produced on either agar or boiled rice. Little research on mass production of *A. aleyrodis* has been published to date. Work on the efficacy of this fungus has not proceeded much beyond limited glasshouse trials.

4.2.3 *Chytridiomycetes* (Coelomomyces *spp.*)

The chytrid parasites of mosquito larvae have received much attention in recent efforts to identify biological controls of mosquitoes (Couch & Bland, 1986). Two facts are prominent: all the known species require copepod hosts, in addition to mosquito larvae, to complete their life cycles; and none has yet been cultured *in vitro*.

4.2.4 *Deuteromycetes*

4.2.4.1 *Beauveria bassiana*. This species has received more developmental attention than any of the other entomopathogenic fungi. Insects that have succumbed to *Beauveria* have a dusty white appearance, giving rise to the name, 'white muscardine'. The infectious unit is a conidium, which germinates on an insect's cuticle, producing a penetration hypha. This hypha penetrates the cuticle by means of mechanical pressure and a mixture of enzymes. Once the haemocoel is invaded, the fungus spreads through the insect as mycelium, or (less often) as free blastospores, and the host is killed within 5–7 days after infection. For details of the biology of *Beauveria* the reader is referred to Steinhaus (1949); some recent reviews of its use can be found in Roberts & Yendol (1981) and Roberts *et al.* (1983).

A key point to remember is that *Beauveria* is dimorphic, growing either by mycelium or yeast-like blastospores *in vitro*, depending upon as yet imperfectly understood factors. Conidia have been the preferred product because of their greater stability and infectivity to insects.

Early mass production attempts used solid substrate fermentation on various substrates: corn meal, wheat bran, rice (Bartlett & LeFebre, 1934; McCoy & Carver, 1941; Dresner, 1949; York, 1958). These early efforts typically involved inoculation of moistened, autoclaved wheat bran with conidia of *Beauveria* from agar slants, then incubating the culture for 6 to 8 days at room temperature. The finished fermentation was dried intact, then pulverised and sifted through a 30-mesh screen. The early field applications of *B bassiana* were made as dust or, less commonly, as an aqueous suspension sprayed on crops.

An alternative method, albeit of limited scope, was to grow *B. bassiana* in sterilised soybean mash spread into petri dishes. This yielded 210 g of conidia (approximately 2×10^{13}) from 461 plates (Beall *et al.*, 1939). Conidia were collected from the surface of the media by scraping and 300–400 petri dishes were required for limited field trials.

Considerable advances have been made since then, notably by workers in the USSR and Czechoslovakia. The Soviets have four production methods for *B. bassiana*, or Boverin ®:

(1)   solid substrates such as heat-sterilised grains;
(2)   solid substrates in a rotating drum;
(3)   a two-phase system in which abundant mycelium is produced in deep tank fermentation, then allowed to sporulate in shallow, open trays;
(4)   submerged liquid fermentation using specific substrates and environ-

mental conditions to obtain conidia rather than the blastospores usually produced in liquid fermentation.

The last method may be specific to certain isolates of *B. bassiana*; we have not found the process to work with other isolates. According to current information the third method is the most preferable and has been scaled up to commercial levels. In 1977, 22 tons of Boverin were produced (Roberts & Yendol, 1981).

Czechoslovakian workers have developed a unique method of conidial production. They employ liquid surface culture within large, inflated plastic bags (Kybal and Vilcek, 1976; Samsinakova *et al.*, 1981). When sporulation is complete, the *Beauveria* is harvested by draining off the liquid, then drying and milling the aerial mycelium and conidia. This method has yielded about 1.4×10^{12} to 10^{15} conidia per square metre of liquid surface.

Mass production of *B. bassiana* has been widely practised in the People's Republic of China: spores of the fungus are produced under relatively primitive conditions on communes (National Academy of Sciences, 1977; Hussey & Tinsley, 1981). There were about 1000 *Beauveria* production units in the PRC by the late 1970s.

Substrates include steamed wheat bran, rice powder, compost, or ground corn stalks. Typically, seed cultures made in 500-ml glass bottles of substrate are used to inoculate 5-kg lots of steamed grain. After 3 days' fermentation, the contents of each lot are mixed with ten times the amount of wheat bran. The fermentation is carried out in flat trays, glass crocks or shallow outdoor pits. After an additional fermentation of 3–4 days, the material is sun-dried and packaged without additional formulation. Typical daily yield on a commune is 50 kg at $8-11 \times 10^7$ conidia per gram or $4.5-5 \times 10^{12}$ total by a staff of seven. Hsu *et al.* (1973) reported yields of $4.6-9.4 \times 10^9$ per gram with a similar production system. All the Chinese methods are very labour-intensive, and thus may not be practical in a Western industrial setting where labour is expensive.

In the US there has been less commercial interest despite intense activity by federal, state and university entomologists exploiting the fungus to control a number of insects. At present, only Abbott Laboratories is studying the development of *B. bassiana* as a biological insecticide for several insect pests. Commercial scale-up of *Beauveria* production by Abbott is underway, using solid substrate, to produce either a wettable powder or a granular formulation.

4.2.4.2 *Beauveria brongniartii (tenella)*. This species, closely related to *B. bassiana*, has been intensively studied by French workers for the control of the cockchafer, *Melolontha melolontha* (Ferron, 1974). This fungus is more host-specific than *B. bassiana* and its blastospore formation is more easily induced in liquid culture. Most efforts have been directed towards blasto-spore production in deep tank fermentation, even though these spores are neither as infectious nor as persistent in the environment as are true conidia.

4.2.4.3 *Metarhizium anisopliae*. This species is commonly called 'green muscardine' because of the green colour of the sporulated mycelium on cadavers. The biology of this species is very similar to that of *Beauveria*. For more details of *M. anisopliae* biology the reader is referred to Ferron (1981) and Steinhaus (1949).

Mass production of *M. anisopliae* by Metchnikoff in 1880, and by Krassilstchik four years later, marked the first use of artificial substrates for growing entomogenous fungi. Krassilstchik (1888) used sterile beer mash injected into large metal containers through a closed network of pipes. Inoculation of the medium was by the residual *Metarhizium* in the containers from the previous fermentation. After 15 days, 180–220 g of spores (approximately 2×10^{13}) were produced from 1 m^3 medium and 55 kg of *Metarhizium* spores were obtained in four months.

This species was generally ignored until the 1960s, when several concerns established small manufacturing facilities in Brazil with assistance from the Boyce Thompson Institute in the US.

Brazilian mass production involves use of autoclaved rice or wheat bran in autoclavable plastic bags (Aquino *et al.*, 1975, 1977). The medium is aseptically inoculated with a blastospore suspension previously produced in liquid shake culture. The fungus grows throughout the substrate as mycelium and, after sporulation, the contents of the bags are dried and milled. Marques *et al.* (1981) described a commercial facility designed to produce 540 kg of *Metarhizium* per year.

Metarhizium has been produced in the USSR in a two-phase system (Goral & Lappa, 1973). The first phase involved submerged fermentation using corn extract and molasses with continuous aeration. After 40–50 h the resulting mycelial mass was transferred into open pans and incubated a further 6 days under controlled conditions. Yields were 5–7.5×10^{12} per kilogram of dried preparation (1–1.5×10^{14} per fermenter batch of 100 litres).

Another approach has been taken by entomologists in Australia. There, a granular bait formulation of *M. anisopliae* was prepared by a two-phase method (D. Pinnock, personal communication). Between 10 000 and 12 000 litres of liquid medium ('agricultural wastage', sucrose, calcium and water) were inoculated with a 10-litre seed of *Metarhizium* blastospores and incubated for 5 days in a standard industrial fermenter. Agitation was by vigorous aeration. The fermenter contents were then mixed with 12 tons of dry bran, hulls, etc., and the mixture was dispensed into plastic-lined sacs and incubated another 5 days. During this period the fungus sporulated intensively to give a final product yield of 4×10^{12} conidia per kilogram of dry bait for a total of 4.36×10^{16} conidia.

4.2.4.4 *Paecilomyces spp.*

Two species of *Paecilomyces*, *P. farinosus* and *P. fumoso-roseus*, have shown some utility in the control of the Colorado potato beetle in Poland (Bajan *et al.*, 1975). Strains of both species are also being evaluated for the control of the brown planthopper in rice (Humphreys *et al.*, 1986). Both species have a biology very similar to that of *B. bassiana* and have been produced on a pilot scale for field evaluation by fermentation on wheat grains. Harvest was accomplished by merely drying the sporulated fungus and substrate and then applying the resulting granules directly to the soil.

Paecilomyces readily produces blastospores in submerged culture. Material has been grown in 5-litre fermenters, and factors affecting yields have only now begun to be studied (Humphreys *et al.*, 1986). The blastospores are infectious to insects, but they suffer the poor stability of other Deuteromycetes.

4.2.4.5 *Verticillium lecanii.*

V. (Cephalosporium) lecanii is a common pathogen of scales and aphids in the tropics and sub-tropics. It is probably a complex of many closely related species (Hall, 1981). Much recent effort was expended by Microbial Resources Ltd (MRL) on commercialising this fungus in the UK for use on glasshouse crops.

Like *Beauveria* and *Metarhizium*, *V. lecanii* is dimorphic. Conidia can be produced from vegetative mycelium on granular solids, e.g. grain, and blastospores are produced in liquid media. Blastospore yields in liquid fermentation can reach $0.7–5 \times 10^{12}$ per litre in dextrose–yeast extract medium (Latgé *et al.*, 1986).

V. lecanii was commercialised in the UK by MRL as Mycotal® for control of aphids and Vertalec® for whitefly. The two products are different strains of *V. lecanii*. Production is based on submerged liquid fermentation yielding blastospores, which are then formulated and which must be stored under refrigeration; storage life is only a few months.

4.2.4.6 *Hirsutella thompsonii*

This pathogen of eriophyid mites has been under study for about 30 years and was commercialised as Mycar® by Abbott Laboratories in the 1970s. *H. thompsonii* has frequently been seen in epizootic proportions among citrus rust mites in Florida, and it was this insect that became the chief target of *H. thompsonii* development. Bermuda grass stunt mite and blueberry bud mite are other target insects. A review of this species is provided by McCoy & Couch (1981).

Both semi-solid and submerged fermentation have been used to mass produce the fungus (McCoy *et al.*, 1975), although Abbott Laboratories employed submerged fermentation followed by aerobic pan fermentation to produce conidia and viable mycelial fragments (McCoy & Couch, 1981). The end product was dried, milled, and packaged as a wettable powder having 9.8×10^9 colony forming units (CFU) per gram. However, mainly because of its short shelf-life at room temperature and stringent high humidity requirements for survival of the fungus on vegetation, Mycar is no longer produced.

4.2.4.7 *Nomuraea rileyi.* This fungus is a pathogen of cabbage looper, soybean looper, velvetbean caterpillar, and *Heliothis* spp. Its biology is similar to that of other Deuteromycetes and its status has been reviewed by Ignoffo (1981).

   N. rileyi has been mass produced on solid media (agar) in petri dishes, yielding 6.3×10^8 conidia per square centimetre of surface, or 4.7×10^{11} per litre agar (Bell, 1975), and on agar media in trays, with about the same yield but at greater efficiency (Ignoffo *et al.*, 1981). Submerged fermentation gave higher yields — about 7×10^{11} per litre of medium — but these were blastospores, not true conidia (Bell, 1975). Blastospores are not infective.

4.2.4.8 *Culicinomyces clavosporus* and *Tolypocladium cylindrosporum.* Both these fungi are specific pathogens of mosquito larvae. *Culicinomyces clavosporus* has been studied extensively by Sweeney and his co-workers in Australia (see Federici, 1981, for an overview of current knowledge). *Tolypocladium cylindrosporum* has only recently been identified (Soares *et al.*, 1979) and is still under preliminary evaluation as a mosquito bio-control. Current data on efficacy indicate that at least 10^{13} *Culicinomyces* or *Tolypocladium* conidia per hectare of mosquito habitat are needed for effective control. Both conidia and blastospores of *T. cylindrosporum* are infectious.

   Spores of both fungi have been produced on wheat-bran substrate, in shake flasks of liquid media or in small fermenters. Liquid shake flask culture of *Culicinomyces*, using corn meal extract, corn steep liquor, or standard nutrient broths, yields $1-2 \times 10^{10}$ conidia per litre after 7–10 days. It has been grown in 750–1 litre seed fermenters in Australia (A. Sweeney, personal communication). *Tolypocladium* produced 1×10^{11} blastospores per litre of Sabouraud broth in 4–6 days (Pillai, 1982). Both species need considerably more evaluation before their potential can be determined.

4.3 Fundamentals of solid substrate fermentation

From the previous section it is evident that a wide variety of methods have been used with greater or lesser success to produce mycoinsecticides. Each organism has its own particular requirements — substrate type, nutrients, etc. — and mass production efforts must be individually tailored for optimal yields. Solid substrate fermentation (SSF) does have a major advantage in allowing the fungus to sporulate naturally, i.e. via aerial mycelium. For this reason, solid substrates rather than submerged systems have been the methods of choice with Deuteromycetes and they need to be further scrutinised for potential in the commercial-scale production of mycoinsecticides.

Solid substrate fermentation is actually a dense slurry system in the absence, or near-absence, of free liquid in which water-insoluble materials are used for microbial growth (Moo-Young *et al.*, 1983). In this respect, the processes used to produce fermented foods (Hesseltine, 1965, 1977b; Hesseltine *et al.*, 1976; Wang & Hesseltine, 1979; Aidoo *et al.*, 1982; Steinkraus, 1983) and fungal metabolites (Hesseltine, 1972, 1977a), as well as for bioconversion of plant, animal and domestic wastes into useful products (Aidoo *et al.*, 1982; Moo-Young *et al.*, 1983), have been solid substrate fermentations (Martinelli & Hesseltine, 1964).

A measure of the importance of such processes is shown by the fact that the koji process alone was used in Japan to produce 1 200 000 kilolitres of soya in 1974, 1 800 000 kilolitres of sake in 1973, and 587 000 tons of miso in 1974 (Yamada 1977).

Most of the basic studies on SSF have been done with a view to the above-mentioned processes, not for production of entomopathogens. For example, neither of two highly regarded reviews of SSF (Moo-Young *et al.*, 1983; Mudgett, 1986) even mentions entomopathogens. Nevertheless, publications on composting, enzyme production, and fermentation of oriental foods are of considerable importance. A great deal of research has been done on these processes, much of it on a large scale and applicable to mycoinsecticide production. And understanding of the basic principles will expedite achievement of the yield increases needed to commercialise fungi as mycoinsecticides.

4.3.1 *Substrates*
Solid substrate fermentation processes have been classified according to whether or not the substrate had nutritive value, as shown in Table 4.2. A list of potential substrates would be virtually endless. For example, almost any organic waste can be used — and usually has been — at one time or another.

Inert substrates, especially, possess significant advantages: because a defined medium can be added to an inert substrate, there need be no concern about the interplay of complex carbon sources. Similarly, the effects of salts, vitamins and other additives are more easily interpreted. In this regard, Miller & Churchill (1986) have tabulated a great number of

Table 4.2. Solid substrates used in fermentation

Inert (non-nutritive)[a]		
Paper	Vermiculite	Attapulgite
Perlite	Wood	
Nutritive[b]		
Corn	Soybeans	Sorghum
Buckwheat	Wheat	Millet
Oats	Rice	Coffee pulp
Potatoes	Barley	
Peanuts	Straw	

[a] With a number of fungi, paper and wood would be considered nutritive.
[b] Many nutritive substrates may be used whole, dehulled, pearled, cracked, or as meal. Brans of these materials may also be used.

carbon and nitrogen sources and their ingredient analyses, most of which can be used in SSF.

On the other hand, economics frequently dictate that substrates include complex nutrients. Not only is the cost often lower but also the complex nutrient mixtures give better growth. Even with these substrates, however, there is a frequent need for additional nutrients. Soybeans, for example, contain little readily available carbohydrate and should be mixed with wheat, which has a high starch content. Mudgett (1986) cites a number of instances where addition of monosaccharide stimulated growth on woods high in lignin. Maheva *et al.* (1984) suggested that some starchy substances may have to be supplemented with mineral salts to improve the buffering capacity of the medium and spore yield.

Substrate particle size and related physical factors are important. Hesseltine (1972) pointed out that substrates must be within a limited size range for optimal results. Corn or soybeans, for instance, should be cracked into five or six pieces, but whole rice kernels or pearled wheat were said to be ideal in size. Small particle sizes provide a high surface area to volume ratio, which leads to better nutrient absorption, gas exchange and heat transfer. However, smaller particles also pack closer together, thus reducing gas exchange and heat transfer. Compromise among these opposing factors can lead to the selection of different sized particles for different fungi. Such selection may be easier with inert substrates, but it is also possible to modify nutritive substrates which can be chopped, husked, cracked, crushed or ground to an optimal size.

Treatments may also be required to facilitate attachment of the inoculum to the substrate and to ensure mycelial penetration into particles. Steinkraus (1983) cites several examples of oriental foods where rice must be hulled and milled, soaked in water, and steamed before being used.

This treatment softens the rice and renders it more permeable to the fungus.

After physical and/or chemical treatment of the substrate, it is important that the physical integrity be maintained otherwise aeration and heat transfer will suffer. For this reason, Maheva *et al.* (1984) believed buckwheat to be superior to 12 other cereal and starchy substrates, because it retained its structure and did not agglomerate.

4.3.2 *Moisture*

While water is necessary in SSF, it exists in an absorbed or complexed form within the solid matrix. Hesseltine (1965) pointed out that a successful solid state fermentation depends on having the moisture content high enough for fungal growth but not so high as to encourage bacterial growth. Tengerdy (1985) emphasised that single-cell organisms like bacteria always require free water whereas filamentous fungi can grow in the absence of free water. At least 12% moisture must be present before fungi will grow (Hesseltine, 1972), but free water, depending upon the substrate, does not occur until the moisture level reaches 40–80% (Cannel and Moo-Young, 1980). The upper limit for solid state systems is a function of absorbency. Free water becomes apparent in maple bark at 40% moisture but in wheat straw it only becomes noticeable at about 75% moisture (Cannel and Moo-Young, 1980). For most solid substrates free liquid becomes apparent above 80% moisture (Moo-Young *et al.*, 1983). Hesseltine (1972) felt that optimum results were obtained at about 28% moisture.

Excessive moisture may also cause the substrate particles to stick together. Tengerdy (1985) described a properly moistened substrate as one which had 'a surface film of water to facilitate dissolution and mass transfer of nutrients and oxygen, but interparticle channels would be left free to permit oxygen diffusion and heat dissipation'.

4.3.3 *Temperature*

Given the importance of the effect of temperature on fungal growth and sporulation, it is surprising that it has been stressed so little in the production of mycoinsecticides with solid substrates. Certainly, submerged fermentations are known to require extensive cooling to overcome the heat generated by cellular metabolism. Mudgett (1986) pointed out that, for the same reason, heat will be generated within a solid substrate. Several koji processes have generated up to 600 kcal per gram of solid substrate (Aidoo *et al.*, 1982).

This heat generation can cause rapid temperature increases in the substrate mass where heat transfer is limited. Takamine (1914) cited an *Aspergillus* fermentation which was initially at 30°C but rose to 42°C within 18 h as fungal growth proceeded. Another report (Hao *et al.*, 1943) confirmed the temperature rise from 30 to 45°C when *Aspergillus, Penicillium* or *Rhizopus* were grown on wheat bran. In the latter case the temperature would have risen further except for the cooling effect of aeration. Rathbun and Shuler (1983) also presented data for *Rhizopus oligosporus*, which showed over a 10°C increase during 34 h in a 25-cm-deep tempeh fermentation. Temperature increase was felt to be directly related to the thickness of the substrate as well as to the metabolic activity of the culture (Hayes, 1977, as cited in Maheva *et al.*, 1984). With a 5-cm thickness of bran substrate, good mould growth occurred only in the outer 2.5 cm. Temperature was so high inside the substrate mass that the initial fungal growth was killed (Underkofler *et al.*, 1947). Although many of the entomogenous fungi grow more slowly than the above-mentioned species, heat generation can still be a significant problem and may have to be taken into consideration during mass production.

The low moisture prevalent in SSF also creates especially difficult conditions for heat transfer (Moo-Young *et al.*, 1983). Two closely related conditions have significant bearing on the potential temperature rise: moisture content and rate of aeration. Tengerdy (1985) stated that moisture was an influence because thermal diffusion depended directly on substrate moisture content, and because water evaporation was the greatest source of cooling available. Although heat build-up was believed to be the greatest problem in scaling up solid substrate fermentations, controlled evaporation with continuous water replacement should improve heat transfer and vegetative growth. If heat build-up were then allowed to cause the evaporation of most of the water, vegetative growth would stop and sporulation would be stimulated.

Moo-Young *et al.* (1983) emphasised that heat transfer in SSF was closely related to aeration. With the normally low moisture content, heat transfer is inhibited. Aeration rate must therefore be increased to decrease substrate temperature.

4.3.4 *Aeration and agitation*

Aeration plays a multifaceted role in SSF. Not only is it required for gas transfer, but, as previously mentioned, it also aids heat and moisture exchange between substrate and atmosphere. Mechanical mixing of the substrate is usually not possible because it might break up mycelium, slow down growth and inhibit sporulation (Underkofler *et al.*, 1939; Bajracharya & Mudgett, 1980). Silman (1980) felt that, in a small rotary fermenter, mechanical scraping action on the mycelium prevented spore formation of *Aspergillus awamori* by breaking off the conidiophores, the swollen vesicle or the phialides. A helical screw fermenter was developed to overcome this problem. It reportedly gave thorough mixing without damage to the mycelium (Tengerdy, 1985).

Because of mixing problems, obligate aerobes must obtain oxygen and discharge carbon dioxide under relatively stagnant conditions (Mudgett, 1980). With some fungi, such as *Chaetomium cellulyticum* (Ulmer *et al.*, 1981) and *Aspergillus oryzae* (Bajracharya & Mudgett, 1980), growth and enzyme production were inhibited by high levels of carbon dioxide and stimulated by high oxygen levels. Silman (1980) reported that high aeration levels promoted sporulation of *A. awamori*. This was confirmed for tempeh fermentation by Wang and Hesseltine (1979). Hesseltine (1972) pointed out that *A. parasiticus* sporulated heavily on rice in stationary culture but not when placed in a rotary shaker. Mudgett (1980) suggested the use of closed systems with oxygen-enriched air and reduced carbon dioxide to increase yields.

4.3.5 *Hydrogen ion concentration (pH)*
Underkofler (1939, 1947) moistened wheat bran with 0.1 to 0.3 N hydrochloric acid to a pH of 3.5 to 4.5, which helped to reduce the extent of steam sterilisation needed to eliminate contaminants. The acidity of the medium inhibited growth of bacteria. Steinkraus (1983) has also reported that acidification is essential to prevent the growth of spoiling microorganisms in sake fermentation. In recent industrial practice, lactic acid is added to the mash to reduce pH to 3.6–3.8.

4.3.6 *Equipment for fermentations*
Through the centures of history of solid substrate fermentation, a great variety of equipment has evolved (Moo-Young *et al.*, 1983). Aidoo *et al.* (1982) presented an extensive list of relevant vessel types and processes. For the purpose of this review, only equipment that might have application to the production of mycoinsecticides will be covered i.e. bags, trays and rotating drums.

4.3.6.1 *Bags.* Lotong and Suwanarit (1983) described the production of koji spore inoculum in polypropylene bags. The 16×25 cm bags containing 10g of rice were inoculated with *Aspergillus flavus* or *A. oryzae* and incubated at 30°C. Yields were 10^9 per gram. After drying, the spores were stable for at least three months at 28–35°C. The operation was scaled up to use 40×60 cm bags and is now used by two local Thai factories to produce koji inoculum. Their process appears to be very similar to that used by Aquino *et al.* (1977) to produce *Metarhizium*.

Although the use of small bags as the basis of a production operation may seem impractical, the technique has the advantage of being applicable to areas where elaborate support systems are not available. Scale-up potential would appear to be a problem because aeration and heat transfer would soon be limiting as bag size increases.

4.3.6.2 *Trays*. The koji process is basic to most tray operations. This method is used to make a variety of foods in the Orient and has typically been accomplished by spreading treated rice in a 5- to 8 cm-deep layer in woven bamboo baskets or wooden trays. These containers are then stacked vertically with 10 cm gaps between them for air circulation (Aidoo *et al.*, 1982; Moo-Young *et al.*, 1983). Large incubators with cooling water circulation systems and forced humidified aeration have been introduced in modern Japan (Aidoo *et al.*, 1982).

Takamine (1914) adapted the koji process to grow *Aspergillus oryzae* on wheat bran to produce amylase. Trays with bottoms of wire netting fine enough to hold the bran particles were filled 3.75 cm deep and stacked 5 cm apart. Air, supersaturated with water, was blown into the chamber until sporulation started, at which time dry air was substituted to dry the substrate.

Hao *et al.* (1943) prepared mould amylases with *Aspergillus*, *Mucor*, *Penicillium* and *Rhizopus* in 3-litre aluminium pots 18 cm in diameter and 11 cm high. The pot bottoms were perforated with 3.2 mm holes. Seven hundred and fifty grams of acidified bran were placed in each pot and the pot was then placed on a layer of cotton batten. After 8 h, when the temperature had risen to 40°C from the initial 30°C, air was passed through the bran to keep the temperature below 45°C. The direction of the air flow was reversed at hourly intervals by alternate application of pressure and vacuum. Further work with *Aspergillus* (Underkofler, 1947) used galvanised iron pans, 61 cm by 89 cm and 10 cm deep, with an air inlet located in the centre and outlets at each corner. After inoculation with spores, a 6 cm layer of bran in the tray was aerated with 1200–1800 ml air per minute. Heavy sporulation occurred after 4–5 days.

Uniformly good mould growth was also achieved with unaerated 30 cm-square trays 5 cm deep. When depth was increased to 30 cm and the temperature was controlled by forced air circulation, growth continued to be good. This process was scaled up to a 91 × 91 × 30 cm deep box with air being forced horizontally through the 30 cm thicknesses of substrate. Substrate thicknesses greater than 91 cm were not successful owing to temperature build-up and packing of the bran. In another unit, containing 255 litres of bran and inclined 45°, substrate packing was not serious. Air requirements for 36 h of incubation averaged 0.96 cubic metres per minute (CMM) per 0.03 cubic metres of bran with a maximum of 1.56 CMM at the peak heating period. Maximum substrate volume was 3.66 × 3.66 × 1.22 m with an initial moisture content of 51% (w/w). If the bran dried too fast, sporulation was poor; if drying was too slow, autolysis occurred with rapid loss of spore viability.

Pamment *et al.* (1978) used stainless steel trays 40 × 32 × 8 cm, equipped with lids and false bottoms of perforated stainless steel. Sterile humidified air was passed through a 1 cm layer of sawdust inoculated with *Chaetomium cellulyticum*. In 20 days, mycelium permeated through the sawdust but was greatest at the top of bottom surfaces. Sporulation occurred after 5 days.

4.3.6.3 *Rotary fermenters.* Takamine (1914) used a plain iron drum holding up to 4800 lb (2177 kg) of bran to prepare amylase. The drum had an air inlet on one side with an exit on the other. The drum rotated at 1 rpm and, with a charge of 1090 kg, used 1600–1800 CFM humidified air to keep the temperature within limits.

Underkofler (1939) employed a similar technique with iron 19-litre cans containing 1–2 kg bran. The cans were aerated and rotated for 20 min every 2 h until the spore inoculum germinated. After germination, the cans were rotated at 1 rpm for 40–50 h.

Although rotary fermenters worked well for enzyme production, the movement resulted in poor spore production. Despite this defect, there is still interest in the possible use of rotary fermenters because of their potential for scale-up. Based on our own experience and reports in the literature, it is difficult to imagine that this approach will produce spores of commercially attractive entomopathogenic fungi in satisfactory quantities.

4.4 Technical problems in mass production

Regardless of the type of process and equipment, the recovery method may be all important for successful mass production of a mycoinsecticide. (As in personal finance, it's not what you make but what you keep.)

Harvesting from surface cultures on agar is straightforward. A number of vacuum collection systems have been designed (Bartlett & LeFebre, 1934; Aquino *et al.*, 1977). The spores are either directly vacuumed from the surface or scraped off and then collected with a vacuum system. Spore harvest from solid substrates is usually accomplished by drying the culture, then pulverising the mass and formulating the product as a dust or granule. Alternatively, a dry or partially dry culture can be tumbled to knock off the spores, which are then aspirated into a collection vessel.

Rapid drying at temperatures below 30–35°C seems critical to preservation of spore viability. Slow drying over a number of days often results in low initial viabilities and poor shelf-life whereas high temperatures cause rapid death of conidia. Thus, milling can cause loss of viability as a result of heat build-up rather than from abrasion.

Harvesting from submerged fermentations is much easier because the spores can be readily collected and concentrated by centrifugation, filtration, etc., then dried. Submerged liquid fermentations also have a number of highly desirable attributes: they are cheaper, particularly on a very large scale; environmental factors (pH, pO_2, pCO_2, nutrient levels) are more easily controlled, and growth is easily monitored. Furthermore, submerged fermentation fits much more easily into the framework of current industrial microbiology, especially in the West.

While submerged fermentation may seem to have the greatest potential for mass production, there are significant problems when the major Deuteromycetes are grown in submerged systems. Most of the entomogenous Deuteromycetes are dimorphic: they can grow either by mycelial proliferation or by budding of blastospores. The factors controlling the development of each of these growth forms are not yet fully understood. *Metarhizium*, *Beauveria*, *Hirsutella*, *Nomuraea* and *Verticillium* produce blastospores rather than conidia in submerged culture, often with high yields (10^{11} per litre) within 72–96 h. These blastospores are very sensitive to drying, much less persistent in the environment, and, except for *Verticillium*, much less infectious than are conidia (Bell, 1975; Hall, 1981; S.T. Jaronski, unpublished data).

It must be remembered that the end product of entomopathogen production is a living entity sensitive to environmental extremes. Its viability must be sustained, not only during harvest and formulation but also during some period of storage. Ideally, a mycoinsecticide should remain completely viable for at least 12 months at 25–30°C. Harvesting methods may not kill the spores outright but may still stress them sufficiently to greatly reduce shelf-life. To date, only *B. bassiana* and *M. anisopliae* have satisfied these requirements (Ward & Roberts, 1981; Daoust *et al.*, 1983; S.T. Jaronski & J. Nyberg, unpublished data).

The Soviets seem to have made significant advances in inducing *B. bassiana* to produce conidia rather than blastospores in submerged fermentation, by a combination of specific nutrients and aeration conditions (Goral, 1973). their success, however, may be limited to certain isolates of the fungus more prone to conidiation than others. Our attempts to produce conidia from other isolates of *Beauveria* using their methods have failed.

Thus, a number of alternative systems — two-phase, liquid-pillow, etc. — have arisen as compromise techniques. Most are, however, primitive and of limited scale, certainly inadequate for commercial demands in a large crop such as corn. Relatively little has been learned from fermentation microbiology particularly with regard to solid substrate fermentation, which may (especially for the Deuteromycetes) prove to be the most practical method.

Mass production can be further confounded by the need to produce several strains of a particular fungus, each specific for a given target insect. Host specificity demonstrated by species, or even strains within a species, is universal among the entomogenous fungi. This variability is also reflected in differences in growth, sporulation, and nutritional requirements. We

have frequently found that isolates desirable from the standpoint of pathogenicity are poor growers or sporulators and vice versa, or that a substrate
excellent for one isolate is poor for others.

4.5 The key problem: sufficient production

The fundamental question underlying commercial production is whether
enough infectious bodies (conidia, blastospores, etc.) can be produced to
effectively control a target insect over an area sufficiently large to attract
serious commercial interest. This is the key to practical feasibility.

Typical use rates for the most developed of the entomogenous fungi are
given in Table 4.3. Notice that few are below 1×10^{13} per hectare, and
none is below 5×10^{12}. When that rate is extrapolated to commercially
attractive hectarage, the amount of production needed can be prodigious.

Let us consider an example: the use of *B. bassiana* as a mycoinsecticide
for the control of European corn borer, *Ostrinia nubilalis*. A preliminary
assumption is a large attractive market for industry: corn, which involves 5
million ha in the state of Illinois alone. With an anticipated mature market
share of 20%, *Beauveria* would have to be produced for 1 million ha.

Large-scale field tests of *Beauveria* as dust and granular formulations in
the People's Republic of China revealed that effective control was achieved
by $1-6 \times 10^{13}$ conidia per hectare (Hsu *et al.*, 1973). If that rate was
applied to the market share of Illinois corn, a total of $1-6 \times 10^{19}$ conidia
would be needed.

Published yields of *Beauveria* conidia from liquid surface culture are
typically 1×10^{14} per square metre (J. Weiser, personal communication); 3
$\times 10^{11}$ per litre for submerged liquid fermentation (Goral, 1973), and

Table 4.3.   Typical use effective rates of entomopathogenic fungi

Fungus	*Insect*	*Rate pcs hectase*
Beauveria bassiana	*Lepidoptera* on cabbage	$>9 \times 10^{14}$
	Ostrinia nubilalis	$1 \times 10^{13}-6 \times 10^{13}$
	Leptinotarsa decemlineata	$6 \times 10^{13}-1.2 \times 10^{14}$
	Sitona sp.	$1.2 \times 10^{16}-1.2 \times 10 +$ {17}
Beauveria brongniartii	*Melolontha* spp.	$5 \times 10^{12}-4 \times 10^{14}$
Nomuraea rileyi	*Trichoplusia ni*	1.4×10^{13}
	Heliothis zea	5×10^{13}
Verticillium lecanii	Aphids	$5 \times 10^{12}-6 \times 10^{13}$
Hirsutella thompsonii	*Panonychus citri*	$>5 \times 10^{8}$
Culicinomyces clavosporus	Mosquito larvae	1×10^{13}

approximately 7×10^{12} per kilogram of solid substrate (National Academy of Sciences, 1977). These yields translate into 100 000 to 600 000 m^2 of surface culture; 62 million to 187 million litres of fermentation capacity; or $1.4–8.6 \times 10^3$ metric tons of solid substrate per year.

Another example is the production of *Culicinomyces*. The effective field rates are around 1×10^{10} per square metre. Current submerged fermentation methods yield $1–2 \times 10^{11}$ conidia per litre, which translates into 1000 litres of fermenter capacity per hectare.

Production facilities of this scale present a substantial capital investment, in addition to the requirements of development and registration of a mycoinsecticide. Even with economies of scale, this investment must be considered in the light of the need for pricing that is competitive with existing chemical controls. The implication of such mathematics is that field efficacy and/or efficiency of mass production must be increased by several orders of magnitude to be practical and competitive. Alternatively, application strategies will have to be carefully considered to maximise the efficiency of a mycoinsecticide. The strain specificity of many of the entomogenous fungi complicates the situation: it is possible that different strains will have to be produced for different target insects — a division of resources and fermentation capacity.

Only the Deuteromycetes have been readily mass produced on some scale. The other entomogenous fungi, with their more stringent growth requirements, pose a substantial challenge before they can be considered as practical for commercialisation. Until low cost, uncomplicated mass production can be realised, successful commercialisation of mycoinsecticides will not be feasible. The two most significant challenges before us are (1) submerged fermentation of Deuteromycetes to produce aerial conidia, or a mastery of solid substrate fermentation to efficiently produce the large number of conidia required, and (2) satisfactorily long shelf-life of preparations at room temperature.

References

Aidoo, K.E., Hendry, R. & Wood, B.J.B. (1982). Solid substrate fermentations. *Advances in Applied Microbiology*, 28, 201–37.

Aquino, de L.M.N., Cavalcanti, V.A., Sena, R.C. & Queiroz, G.F. (1975). A new technique for the production of the fungus *Metarhizium anisopliae*. *Boletim Tecnico CODECAP*, 4, 1–31.

Aquino, de L.M.N., vital, A.F., Cavalcanti, V.L.B. & Nascimento, M.G. (1977). Culture of *Metarhizium anisopliae* (Metchn.) Sorokin in polypropylene bags. *Boletim Tecnico CODECAP*, 5, 7–11.

Bajan, C., Kmitova, K. & Wojciechowska, M. (1975). A simple method for obtaining the infectious material of entomogenous fungi. *Bulletin of the Polish Academy of Sciences*, Series II, 23, 45–7.

Bajracharya, R. & Mudgett, R.E. (1980). Effects of controlled gas environments in solid substrate fermentations of rice. *Biotechnology and Bioengineering*, **22**, 2219–35.

Bartlett, K.A. & LeFebre, C.L. (1934). Field experiments with *Beauveria bassiana* (Bals.) Vuill., a fungus attacking the European corn borer. *Journal of Economic Entomology*, **27**, 1147–57.

Beall, G., Stirrett, G. M. & Conners, I.L. (1939). A field experiment on the control of the European Corn Borer, *Pyrausta nubilalis* Hubn., by *Beauveria bassiana* Vuill. *Scientific Agriculture*, **19**,(8), 531–4.

Bell, J.V. (1975). Production and pathogenicity of the fungus, *Spicaria rileyi* from solid and liquid media. *Journal of Invertebrate pathology*, **26**, 129–30.

Cannel, E. & Moo-Young, M. (1980). Solid substrate fermentation systems. *Process Biochemistry*, **15**, 2–7, 24–8.

Couch, J.L. & Bland, C.E. (1986). *The genus Coelomomyces*. Acadmic Press, New York.

Daoust, R.A., Ward, M.A. & Roberts, D.W. (1983). Effect of formulation on the viability of *Metarhizium anisopliae* conidia. *Journal of Invertebrate Pathology*, **41**(2), 151–60.

Domnas, A., Srebro, J.P. & Hicks, R.F. (1977). Sterol requirements in the mosquito-parasitizing fungus, *Lagenidium giganteum. Mycologia*, **69**, 875–86.

Dresner, E. (1949). Culture and use of entomogenous fungi for the control of insect pests. *Boyce Thompson Institute Contributions*, **15**(6), 319–33.

Federici, B.A. (1981). Mosquito control by the fungi, *Culicinomyces, Lagenidium*, and *Coelomomyces*. In: *Microbial Control of Pests and Plant Diseases 1970–1980*, ed. H.D. Burges, pp. 555–72. Academic Press, New York.

Ferron, P. (1974). Essai de lutte microbiologique contre *Melolontha melolontha* par contamination du sol a l'aide de blastospores de *Beauveria tenella. Entomophaga*, **19**, 103–14.

Ferron, P. (1981). Pest control by the fungi *Beauveria* and *Metarhizium*. In: *Microbial Control of Pests and Plant Diseases 1970–1980*, ed. H.D. Burges, pp. 465–482. Academic Press, New York.

Goral, V.M. (1973). The influence of conditions of deep culture on the formation of *Beauveria bassiana* and *tenella* conidia, two entomopathogenic fungi. *Zakhyst Roslyn, Kiev*, **18**, 66–72.

Goral, V.M. & Lappa, N.V. (1973). In-depth and surface culture of the green muscardine fungus. *Zashchita Rastyeni*, **1**, 19.

Hall, R.A. (1981). The fungus *Verticillium lecanii* as a microbial insecticide against aphids and scales. In: *Microbial Control of Pests and Plant Diseases 1970–1980*, ed. H.D. Burges, pp. 483–98. Academic Press, New York.

Hao, L.C., Fulmer, E.I. & Underkofler, L.A. (1943). Fungal amylases as saccharifying agents in the alcoholic fermentation of corn. *Industrial & Engineering Chemistry*, **37**(7), 814–18.

Hesseltine, C.W. (1965). A millenium of fungi, food and fermentation. *Mycologia*, **57**, 149–97.

Hesseltine, C.W. (1972). Solid state fermentation. *Biotechnology and Bioengineering*, **14**, 517–32.

Hesseltine, C.W. (1977a). Solid state fermentation. *Process Biochemistry*, **12**(6), 24–7.

Hesseltine, C.W. (1977b). Solid state fermentation. *Process Biochemistry*, **12**(7), 29–32.

Hesseltine, C.W., Swain, E.W. & Wang, H.L. (1976). Production of fungal spores as inocula for oriental fermented foods. *Developments in Industrial Microbiology*, **17**, 101–15.

Hsu Cheng-fung, Chang Yung, Kuei Cheng-ming, Han Yu-mei, & Wang Hwei-hsien (1973). Field application of *Beauveria bassiana* (Bals.) Vuill. for controlling the European corn borer. *Acta Entomologica Sinica*, **16**(2), 203–6.

Humphreys, A.M., Inch, J.M.M., Gillespie, A.T. & Trinci, A.F.J. (1986). Blastospore production in *Paecilomyces* spp. In: *Fundamental and Applied Aspects of Invertebrate Pathology*, eds Samson, R.A. Vlak, J.M. & Peters, D., p. 239. Foundation of the Fourth International Colloquium of Invertebrate Pathology, Wageningen, The Netherlands.

Hussey, N.W. & Tinsley, T.W. (1981). Impressions of insect pathology in the People's Republic of China. In: *Microbial Control of Pests and Plant Diseases 1970–1980*, ed. H.D. Burges, pp. 785–96. Academic Press, New York.

Ignoffo, C. (1981). The fungus *Nomuraea rileyi* as a microbial insecticide. In: *Microbial Control of Pests and Plant Diseases 1970–1980*, ed. H.D. Burges, pp. 513–38. Academic Press, New York.

Jaronski, S.T. (1982). Oomycetes in mosquito control. Symposium on biological control of vectors, *Proceedings of the Third International Colloquium on Invertebrate Pathology*, 6–10, September 1982. University of Sussex, Brighton, UK, pp. 420–4.

Jaronski, S.T. (1986). Mass production of deuteromycetes for insect control, a critical appraisal. In: *Fundamental and Applied Aspects of Invertebrate Pathology*, eds R.A. Samson, J.M. Vlak, & D. Peters, pp. 653–6. Foundation of the Fourth International Colloquium of Invertebrate Pathology, Wageningen, The Netherlands.

Jaronski, S.T. & Axtell, R.C. (1984). Simplified production system for the fungus *Lagenidium giganteum*, for operational mosquito control. *Mosquito News*, **44**(3), 377–81.

Kerwin, J.L. & Washino, R.K. (1983). Sterol induction of sexual reproduction in *Lagenidium giganteum*. *Experimental Mycology*, **7**, 109–15.

Kerwin, J.L. & Washino, R.K. (1986). Oosporogenesis by *Lagenidium giganteum* in liquid culture. *Journal of Invertebrate Pathology*, **47**, 258–70.

Krassilstchik, I.M. (1888). La production industrielle des parasites vegetaux pour la destruction des insectes nuisables. *Bulletin Scientifique de la France et Belge*, **19**, 461–72.

Kybal, J. & Vilcek, V. (1976). A simple device for stationary cultivation of microorganisms. *Biotechnology and Bioengineering*, **18**, 1713–18.

Latgé, J.P. (1986). The Entomophthorales after the resting spore production stage. In: *Fundamental and Applied Aspects of Invertebrate Pathology*, eds R.A. Samson, J.M. Vlak, & D. Peters, pp. 651–2. Foundation of the Fourth International Colloquium of Invertebrate pathology, Wageningen, the Netherlands.

Latgé, J.P. & Perry, D.B. (1980). Utilization of an *Entomophthora obscura* resting spore preparation in biological control experiments against cereal aphids. *Organization Internationale de Lutte Biologique/Section Regionale Ouest Palearctique, Bulletin III*, **4**, 19–25.

Latgé, J.P., Hall, R.A. Cabrera Cabrera, R.I. & Kerwin, J.C. (1986). Liquid fermentation of entomogenous fungi. In: *Fundamental and Applied Aspects of Invertebrate Pathology*, eds R.A., Samson, J.M. Vlak, & D. Peters, pp. 603–6. Foundation of the Fourth International Colloquium of Invertebrate Pathology, Wageningen, The Netherlands.

Latgé, J.P., Soper, R.S. & Madore, C.D. (1977). Media suitable for industrial production of *Entomophthora virulenta*, zygospores. *Biotechnology and Bioengineering*, **19**, 1269–84.

Lisansky, S.G. & Hall, R.G. (1983). Fungal control of insects. In: *The Filamentous Fungi, Vol. IV, Fungal Technology*, eds J.E. Smith, D.R. Berry & B. Kristiansen, pp. 327–45. Edward Arnold, London.

Lotong, N. & Suwanarit, P. (1983). Production of soy sauce koji mold spore inoculum in plastic bags. *Applied and Environmental Microbiology*, **39**, 430–5.

Maheva, E., Djelveh, G., Larroche, C. & Gros, J.B. (1984). Sporulation in *Penicillum roquefortii*, in solid state fermentation. *Biotechnology Letters*, **6**, 97–102.

Marques, E.J., Vilas Boas, A.M. & Pereira, C.E.F. (1981). Technical guidelines for the production of the entomogenous fungus, *Metarhizium anisopliae* (Metschn.) in laboratories. *Boletim Tecnico PLANALSUCAR*, **3**(2), 5–23.

Martinelli, A. & Hesseltine, C.W. (1964). Tempeh fermentation package and tray fermentations. *Food Technology*, **18**, 167–71.

McCabe, D. & Soper, R.S. (1985). Preparation of an entomopathogenic fungal insect control agent. United States Patent 4,530,834.

McCoy, E.E. & Carver, C.W. (1941). A method of obtaining spores of the fungus *Beauveria bassiana* (Bals.) Vuill. in quantity. *Journal of the New York Entomological Society*, **49**, 205–10.

McCoy, C.W. & Couch, T.L. (1981). *Hirsutella thompsonii*: a potential mycoacaricide. *Developments in Industrial Microbiology*, **20**, 89–96.

McCoy, C.W., Hill, J.A. & Kanavel, R.F. (1975). Large-scale production of the fungal pathogen *Hirsutella thompsonii* in submerged culture and its formulation for application in the field. *Entomophaga*, **20**(3), 229–40.

Miller, T.L. & Churchill, B.W. (1986). Substrates for large-scale fermentations. In: *Manual of Industrial Microbiology and Biotechnology*, eds (A.L. Demain & N.A. Solomon. American Society of Microbiologists, Washington, DC, pp. 122–36.

Moo-Young, M., Moreire, A.I. & Tengerdy, R.P. (1983). Principles of solid state fermentation. In: *The Filamentous Fungi, Vol. IV, Fungal Technology*, eds J.E. Smith, D.R. Berry & B. Kristiansen, pp. 117–44. Edward Arnold, London.

Mudgett, R.E. (1980). Controlled gas environments in industrial fermentations. *Enzyme and Microbial Technology*, **3**, 273–80.

Mudgett, R.E. (1986). Solid state fermentations. In: *Manual of Industrial Microbiology and Biotechnology*, eds A.L. Demain & N.A. Solomon. American Society of Microbiologists, Washington, DC, pp. 66–83.

National Academy of Sciences (1977). Insect control in the People's Republic of China. *Committee on Scholarly Communication with the People's Republic of China Report No. 2*.

Pamment, N., Robinson, C.W., Hilton, J. & Moo-Young, M. (1978). Solid state cultivation of *Chaetomium cellulyticum*, on alkali-pretreated sawdust. *Biotechnology and Bioengineering*, **20**, 1735–44.

Pillai, J.S. (1982). The biology and pathology of imperfect fungi with vector control potential. Symposium on biological control of vectors, *Proceedings of the Third International Colloquium on Invertebrate Pathology*, 6–10 September 1982. University of Sussex, Brighton, UK, pp. 404–8.

Ramakers, P.M.J. & Samson, R.A. (1984). *Aschersonia aleyrodis*, a fungal pathogen of whitefly. II. Application as a biological insecticide in glasshouses. *Zeischrift für angewandt Entomologie*, **97**, 1–8.

Rathbun, B.L. & Shuler, M.L. (1983). Heat and mass transfer effects in static solid substrate fermentations, designs of fermentation chambers. *Biotechnology and Bioengineering*, **25**, 929–38.

Roberts, D.W., LeBrun, R.A. & Semel, M. (1983). Control of the Colorado potato beetle with fungi. In: *Advances in Potato Pest Management*, eds. J.H. Lashomb & R. Casagrande, pp. 119–37. Hutchinson Ross, Stroudsburg, Pa.

Roberts, D.W. & Yendol, W. (1981). Use of fungi for microbial control of insects. In: *Microbial Control of Insects and Mites*, eds H.D. Burges & N.W. Hussey, pp. 125–49. Academic Press, New York.

Samsinakova, A., Kalalova, S., Vlcek, V. & Kybal, J. (1981). Mass production of *Beauveria bassiana* for regulation of *Leptinotarsa decemlineata* populations. *Journal Invertebrate Pathology*, **38**, 169–74.

Silman, R.W. (1980). Enzyme formation during solid substrate fermentation in rotating vessels. *Biotechnology and Bioengineering*, **22**, 411–20.

Soares, G.G., Pinnock, D.E. & Samson, R.A. (1979). *Tolypocladium*, a new fungal pathogen of mosquito larvae with promise for use in microbial control. *California Mosquito and Vector Control Association*, **47**, 51–4. Soper, R.S. (1978). Development of *Entomophthora* species as possible microbial insecticides. *Proceedings, 1st US/USSR Conference on Production, Selection and Standardization of Entomopathogenic Fungi*, Riga. 20–22 May 1977, pp. 270–82.

Soper, R.S. (1985). Role of entomophthoran fungi in aphid control for potato integrated pest management. In: *Advances in Potato Pest Management*, eds J.H. Lashomb & R. Casagrande, pp. 153–77. Hutchinson Ross, Stroudsburg, Pa.

Steinhaus, E.A. (1949). *Principles of Insect Pathology*. McGraw-Hill Book, New York.

Steinkraus, K.H. (1983). Industrial microbiology application of oriental fungal fermentations. In: *The Filamentous Fungi, Vol. IV, Fungal Technology*, eds J.E. Smith, D.R. Berry & B. Kristiansen, pp. 171–89. Edward Arnold, London.

Su, X. -Q., Guzman, D.R. & Axtell, R.C. (1986). Factors affecting storage of mycelial cultures of the mosquito fungal pathogen *Lagenidium giganteum* (Oomycetes; Lagenidiales). *Journal of the American Mosquito Control Association*, `2, 350–4.

Takamine, J. (1914). Enzymes of *Aspergillus oryzae*, and the application of its amyloclastic enzyme to the fermentation industry. *Journal of Industrial and Engineering Chemistry*, **6**(10), 824–8.

Tengerdy, R.P. (1985). Solid substrate fermentation. *Trends in Biotechnology*, **3** (4), 96–9.

Ulmer, D.C., Tengerdy, R.P. & Murphy, R.D. (1981). Solid state fermentation of manure fibres. *Biotechnology and Bioengineering Symposium Series*, **11**, 449–61.

Underkofler, L.A., fulmer, E.I. & Schoen, L. (1939). Saccharification of starchy grain mashes for the alcoholic fermentation industry. *Industrial and Engineering Chemistry*, **31**, 734–8.

Underkofler, L.A., Severson, G.M., Goering, K.J. & Christensen, L.M. (1947). Commercial production and use of mold bran. *Cereal Chemistry*, **24**, 1–22.

Wang, H.L. & Hesseltine, C.W. (1979). Mold modified foods. In: *Microbial Technology, Vol. II*, 2nd edn, eds H.J. Pepper & D. Perlman, pp. 96–131. Academic Press, New York.

Ward, M.A. & Roberts, D.W. (1981). Viability of *Beauveria bassiana* conidia stored with formulation carriers and diluents. *Proceedings XIVth Annual Meeting Society for Invertebrate Pathology*, Montana State University, Bozeman, Montana, 17–21 August 1981. Abstract 39.

Wilding, N. (1981). Pest Control by Entomophthorales. In: *Microbial Control of Pests and Plant Diseases 1970–1980*, ed. H.D. Burges, pp. 539–54. Academic Press, New York.

Yamada, K. (1977). Recent advances in industrial fermentation. *Biotechnology and Bioengineering*, **19**, 1563–1621.

York, G.T. (1958). Field tests with the fungus *Beauveria* sp. for control of the European corn borer. *Iowa State College Journal of Science*, **33**, 123–9.

5 *R. Charudattan*

Inundative control of weeds with indigenous fungal pathogens

5.1 Introduction

The technique of controlling weeds with concerted applications of large doses of inoculum has been regarded as the 'inundative control strategy'

(Daniel *et al.*, 1973; Scheepens & van Zon, 1982; Templeton, 1982; Charudattan, 1984b). As the title of this chapter would imply, the use of indigenous fungal pathogens as inundative weed control agents has lately generated considerable research and commercial interest. In the light of several recent reviews on this subject (Templeton *et al.*, 1979, 1984, 1987; Hasan, 1980; Emge & Templeton, 1981; Charudattan & Walker, 1982; Templeton & Greaves, 1984; Bemmann, 1985; Charudattan, 1985; TeBeest & Templeton, 1985; Mortensen, 1986) and in the hope of avoiding repetition, I have focused the present discussion on the epidemiological and weed control principles underlying the inundative strategy as well as on the status, problems and prospects of this method of weed control.

5.2 Inundative method versus other biological weed control methods

5.2.1 *Inoculative method*
Epidemiologically, biological weed control using pathogens can be accomplished with the inoculative or the inundative methods. Here the term 'inoculative' is used in a sense similar to the injection of a small sample of a substance into a large body to initiate a massive response. The inoculative method is also called the 'classical' strategy (Templeton, 1982) because of its reliance on the time-tested process of foreign explorations, evaluations, and release of pathogens into areas where weed cntrol is desired. The inoculative strategy works well in pathosystems where the host and pathogen have undergone a spatial separation over time. Hence, introduced weeds that lack a complement of pathogens in their adventive ranges are good targets for inoculative control. Theoretically, a plant that is removed from pathogen pressure loses its resistance to the pathogen over time, and when the pathogen is reintroduced the plant tends to be vulnerable. This mechanism may explain the successful spread of *Puccinia chondrillina* on a susceptible biotype of skeletonweed in Australia, where the genetic diversity of the weed is apparently still preserved to varying degrees on two resistant biotypes (Burdon & Marshall, 1981). There is also evidence, albeit based on small sample size, that the Australian specimens of Italian thistles (*Carduus pycnocephalus*) are more susceptible to an Italian strain, *Puccinia cardui-pycnocephali*, than specimens of the weed from the native region of the rust (Olivieri, 1984). Also noteworthy is the possibility that native weeds are susceptible to previously unencountered exotic pathogens occurring on congeneric weed species.

The inoculative pathogen is merely 'released over', or 'inoculated into', small weed infestations relative to the total infestation. If conditions are favourable, an epidemic ensues and the disease exerts sufficient stress to

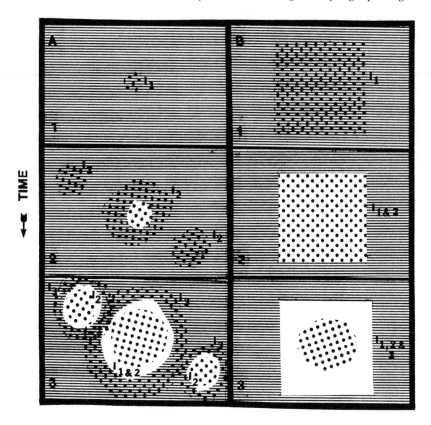

Fig. 5.1. Diagrammatic representation of the differences between (A) inoculative and (B) inundative methods of control in the size of weed infestation, initial infection, disease spread, disease impact on weed population, and temporal aspect of control. Lined areas represent the extent of weed infestation that must be controlled; it is in megahectares in A and in deci- or centihectares in B. Dotted areas denote the amount (severity and spread) of disease. The clear areas within lined areas are the regions where the weed has been controlled or reduced to low levels; the disease persists on these remaining weeds at endemic levels. I_1, I_2 and I_3 are primary, secondary and tertiary foci of infections. In A, the initial focus of infection is small; the area over which control is desired is undefined; and the disease spreads to secondary and tertiary infection foci even as it recedes in the primary focus following the diminishing weed population. In B, the pathogen is applied over a defined area; the secondary and tertiary infections, if any, are limited or inconsequential; and the disease recedes when the weed is controlled. The size of the initial infection can be limited to a few plants in A, but is usually over several hectares in B. The level of weed control is unpredictable in A; a predictable level of control can be expected in B in the treated area. Time is in months or years in A and in days or weeks in B; time 1 corresponds to pathogen introduction and time 3 is when a noticeable level of control has been achieved (A) or is the end of the weed season (B).

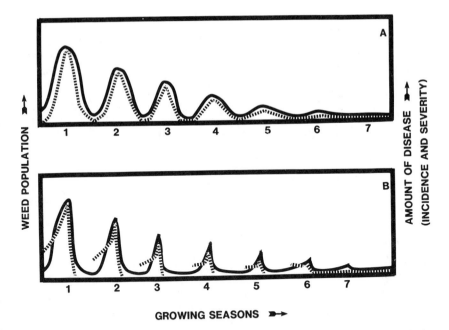

Fig. 5.2. Diagrammatic representation of the differences between (A) inoculative and (B) inundative methods of control in the pattern of weed control. The disease (incomplete line) spreads and reaches severe levels, but the initial disease level is low in A and high in B owing to inundative inoculation. There is gradual increase and decrease of the disease in A; in B the disease increase is rapid, followed by a drastic fall. The seasonal frequency of weed (solid line) is in a symmetrical wave-form in A; the disease increases with increasing weed population, causing control and gradual decline of the population. In B, the weed population builds to a seasonal maximum and the disease reaches a maximum very rapidly, resulting in control and a sudden reduction in weed population accompanied by a fall in disease level. The disease is spread seasonally in A, but in B the seasonal spread is non-existent or inconsequential. After several seasons, the level of weed infestation will be minimal and the disease is likely to reach an endemic status.

control the weed (Fig 5.1 and 5.2). Since this process depends on a gradual increase in disease, the efficacy of the agent also increases slowly, and it may take several months or years to obtain significant weed control. Once released, the agent cannot be stopped without a massive eradication campaign, and it is impossible to predict the success of the agent or to manipulate it to increase or decrease its efficacy.

Because of its relative low cost, the inoculative strategy is preferred for weeds that are distributed over vast areas that yield low or marginal economic returns. The selected pathogens are usually fungi that produce wind-disseminated spores and are capable of rapid sporulation, disease production, and disease spread. Rust fungi such as *Puccinia chondrillina*, utilised for control of skeletonweed, *Chondrilla juncea*, in Australia (Cullen *et al.*, 1973; Hasan & Wapshere, 1973), *Phragmidium violaceum*, used for controlling blackberry, *Rubus constrictus* and *R. ulmifolius*, in Chile (Oehrens & Gonzales, 1974; Oehrens, 1977) and others (Politis *et al.*, 1984; Bruckart & Dowler, 1986; Watson & Clement, 1986) are good candidates for the inoculative method. Facultative saprophytes with wind-disseminated conidia are also suitable for this purpose, as shown by Trujillo (1985), who was able to control the pamakani weed in Hawaii with a *Cercospora*-like fungus imported from the Caribbean. Bacterial and viral pathogens that are disseminated rapidly and widely by vectors can also be used as inoculative agents, but thus far they have been attempted only in the laboratory (see Sands, in Charudattan, 1978; Charudattan *et al.*, 1980).

5.2.2 *Inundative method*

The inundative method consists of applications of massive doses of inoculum to the weed population to create a fast and high level of epidemic when conditions conducive for disease development and weed control are present (Templeton *et al.*, 1979; Templeton, 1982; Charudattan, 1984b, 1985). An inundative agent is manufactured, formulated, standardised, packaged, and registered as a herbicide. It is applied to the weed by conventional chemical appliction methods and tools, and the application schedules are usually integrated with other pesticide schedules (Smith, 1982; Charudattan, 1985; TeBeest & Templeton, 1985; Smith Jr, 1986). Because an inundative agent can produce rapid and complete weed control similar to the fast-acting chemical herbicide, it is the best option for agricultural and intensively managed agroecosystems (Charudattan, 1985). Also, as in the case of the chemical herbicides, there is usually a need for annual applications of the inundative agent since the pathogen generally does not survive in sufficient numbers nor multiply during inter-crop seasons to initiate a fresh epidemic on new weed infestations. Additionally, agricultural soils hold considerable seed reserves to initiate new weed infestations each growing season, making it necessary to reapply the inundative agent. Because of these similarities with chemical herbicides, the inundative tactic has been called the 'mycoherbicide' or the 'microbial herbicide' strategy (Templeton *et al.*, 1979; Templeton, 1982; TeBeest & Templeton, 1985).

Although any pathogen can be applied in a massive inoculum dose, the terms 'inundative agent', and 'inundative control' should be reserved only for a pathogen that can be mass-produced *in vitro* and applied as herbicidal preparations. According to the above definition, obligate parasites such as rusts, smuts and viruses cannot be classified as 'herbicidal' agents since they cannot be cultured in a practical sense and therefore it would not be possible to treat large weed infestations over hundreds or even tens of hectares. Nor can they be used to produce a rapid and high level of weed control within a short time, as is possible with facultative parasites.

5.2.3 *Augmentative method*

It is no doubt possible to inoculate weeds inundatively with obligate parasites using inoculum gathered from naturally infected plants, stock-piled, and applied. Using such techniques, Dyer *et al.* (1982), Phatak *et al.* (1983), and Massion & Lindow (1986) have demonstrated the feasibility of controlling Canada thistle with *Puccinia obtegens*, yellow nutsedge with *P. canaliculata*, and johnsongrass with *Sphacelotheca holci*, respectively. Scheepens & van Zon (1982) have cited similar use of *Puccinia puncti-formis* to control Canada thistle, *Cirsium arvense*, and of cyanophages to control blue-green algae. Unlike the obligate parasites used as classical agents, these pathogens are native or naturalised in their respective regions and occur on native or naturalised weeds, generally causing endemic diseases. But, similar to the inoculative agents, they are capable of self-dissemination and of causing epidemic build-up after an application of inoculum. Some measure of manipulation of the inoculum is also involved, namely, the process of collection, stockpiling, formulation and application. These features set apart the use of native/naturalised obligate parasites against native/naturalised weeds from both the classical and the microbial herbicide strategies. Accordingly, as proposed earlier (Charudattan, 1984b), there is a need to designate a distinct term, the augmentative strategy, for this manipulative inoculation tactic.

In terms of the quantity of inoculum used, the type of inoculum de-ployment, and the management actions necessary to ensure the occurrence of the epidemic, the augmentative strategy can be viewed as being the middle of a continuum between the inoculative and the inundative strate-gies. The amount of inoculum utilised and the area of weed infestation treated are small in the inoculative approach; they are generally large in the case of the inundative method (Figs 5.1 and 5.2). Only limited manage-ment actions are needed for the inoculative method; the augmentative and inundative methods require considerably more management.

The current regulations in the United States governing the use of pathogens as weed control agents also distinguish between inoculative, augmentative and inundative agents (Charudattan, 1982). The inundative agents (microbial herbicides) are regulated and registered under the En-vironmental Protection Agency's biorational pesticide guidelines, whereas the inoculative agents (classical biocontrol agents) are regulated by the US Department of Agriculture (Charudattan, 1982; Bruckart & Dowler, 1986). It is not clear whether the augmentative agents will be subject to EPA and/or USDA regulations; most likely, they will come under USDA and state purviews if interstate movement of the pathogen is involved.

5.3 **Inundative method**

5.3.1 *Principles and general considerations*

Plants in nature are often attacked by many different pathogens, and in the course of evolution the host–pathogen interactions lead to the establishment and maintenance of genetically diverse host biotypes ranging from disease-resistant to tolerant lines (van der Plank, 1963; Burdon, 1982). Similarly, the pathotypes will vary from the mildly virulent to extremely virulent strains. The presence of such variable host and pathogen biotypes ensures the continued coexistence of the plant and the pathogen, and will eventually lead to the persistence of some diseases in nature at low or 'endemic' (as opposed to 'epidemic') levels (Walker, 1969). The endemic level of disease may be sufficient to stress the plant, but it is usually insufficient to produce acceptable levels of weed control (Fig. 5.3), particularly in intensively managed agroecosystems which necessitate rapid and complete weed control.

Endemic disease is almost always present in particular locations at moderate to severe levels; the causative organism survives from year to year in soil, plant debris or host(s); and the environmental conditions are always favourable for disease development on a *microscale*. However, the disease does not become epidemic owing to certain constraints such as host resistance, lack of adequate inoculum, and/or the lack of aggressiveness of the pathogen (Shrum, 1982). Spatial and temporal disjunction between the host and the pathogen may also prevent disease development. Often the

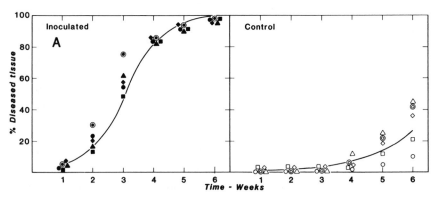

Fig. 5.3.   Epidemic (inoculated) versus endemic (control) levels of leafspot caused by *Cercospora rodmanii* on waterhyacinth. The epidemic condition was produced by inundatively inoculating waterhyacinth plants with the pathogen, whereas the endemic condition occurred on non-inoculated, fungicide-free control plants as a result of natural disease incidence. The disease progressed rapidly under the epidemic and killed a large amount of host tissue. Disease progress was slow under the endemic state and the level of tissue killed was low. The points on the graph represent the levels of disease under different growth rates of the plant (R. Charudattan, unpublished).

discontinuous distribution of weeds limits the spread of the epidemic and causes non-uniform distribution of the inoculum. The quantity of inoculum may be limited by the presence of hyperparasites, poor sporulation or poor inoculum quality (as is common with many rust spores).

On the other hand, epidemic disease occurs periodically over a large population and is generally very damaging to the host. Its onset is dependent on favourable environmental conditions, and is always characterised by a rapid rate of disease progress compared with endemic disease. However, the epidemic situation, characterised by extreme host susceptibility and high degree of pathogen virulence, is not the norm over the long term. What is normal over time is the 'grand cycle of disease' in which the host and the pathogen reach a stage of homeostasis and the disease becomes endemic. An epidemic disease on a weed can therefore be expected to 'run its course' and, over time and space, to become endemic (Fig. 5.2).

Notwithstanding the constraints, a native pathogen can be used to cause an epidemic if massive amounts of its inoculum are applied when conditions are favourable for disease onset. Thus, the endemic condition can be changed to the epidemic by inundative inoculation. However, the successful use of the inundative strategy depends on the creation and maintenance of an effective epidemic (Shrum, 1982), as has been demonstrated on a practical scale by Daniel *et al.* (1973), who successfully controlled northern jointvetch, *Aeschynomene virginica*, in a rice crop with the fungus *Colletotrichum gloeosporioides* f. sp. *aeschynomene*. Subsequently, Ridings *et al.* (1976) also confirmed the feasibility of this approach with a soil-borne pathogen.

The reason for using inoculum as a tool for manipulation is as follows. Among the constraints against disease development (Holcomb, 1982), only inoculum limitation can be easily overcome by man's intervention; little can be done to alter the existing host, pathogen, climatic, temporal or spatial features of the pathosystem. Often inadequate inoculum (i.e. limited quantity or uneven distribution of inoculum) is the leading cause of the lack of an epidemic in nature (Shrum, 1982). Moreover, since indigenous pathogens are already adapted to the ecoclimatic and host conditions in a given region, the provision of abundant inoculum improves the chances for epidemic onset and rapid disease progress.

The inundative use of inoculum makes it possible for man to apply the pathogen when environmental and host conditions ideal for infection, 'infection windows', are present (Shrum, 1982). For instance, in several pathosystems the pathogen is more aggressive on seedlings than on mature plants. This is the case with *Alternaria cassiae*, a pathogen of sicklepod (*Cassia obtusifolia*), which is capable of killing seedlings but loses this ability as the plants mature (Walker & Riley, 1982; Charudattan *et al.*, 1986). Since weeds in crops follow the seasonal patterns of the crops, it is both necessary and easy to control the weeds when they are in seedling stages. With the inundative strategy it is possible not only to time the inoculum application to coincide with the period in which the weed is most susceptible, but also to manipulate other factors such as the provision of humectants or irrigation to counter low moisture conditions in the field. Also, the application can be timed to coincide with the occurrence of conditions ideal for disease development and disease spread.

It is also important to time the inoculation to coincide with the early growth phases of the weed, namely in the spring, for the following reasons. Generally, natural build-up of inoculum is slow during the early phase of the growing season, which results in inoculum deficiency. Since the disease is proportional to inoculum (van der Plank, 1975), the best way to start a destructive disease is to provide additional inoculum when the natural level of inoculum is limiting. By applying an inundative dose of inoculum, the number of infection cycles can be maximised; better advantage can be taken of the favourable environmental conditions for infection; young and susceptible weed tissues can be infected; the negative impact of inoculum isolation can be minimised; and the length of disease cycles and therefore the duration of chronic stress can be increased (Shrum, 1982).

These concepts have been found to be true in the case of the *Cercospora rodmanii* — waterhyacinth pathosystem, in which early season (spring) application of the inoculum, as opposed to the late season (fall) application, results in reductions in weed growth and control (Charudattan, 1984a, 1986a). Spring application interferes with the normal build-up of plant populations and reduces the subsequent weed population, whereas the autumn application may cause plant stress and some mortality (Conway *et al.*, 1979), but as there is no further weed growth the disease has no growth-limiting effect on the weed. In the temperate zone, the contribution of the autumn-applied inoculum towards increasing the amount of initial disease in the following spring will be limited owing to the inoculum-reducing effect of the winter (Leonard, 1982). Furthermore, control attempts must precede seed set in order to be effective; autumn application of inundative agent would generally be ineffective in this respect.

It is generally believed that, owing to the high cost of control, weeds of rangelands, waterways and 'undisturbed' areas are best controlled through inoculative strategy. Since there is no need for 'immediate' control as in crop situations, the time allowable for epidemic development and the duration of disease stress can be fairly long. It is assumed that there is no urgency and no time constraint in these weed control situations. But our experience with the waterhyacinth — *Cercospora rodmanii* system indicates that only the inundative strategy, not the inoculative, is suitable for managing this weed in the waterways of the southeastern United States (Charudattan, 1984a). The reasons for this have been discussed (Charudattan, 1984a) and they include public expectation of total and quick control, the low level of disease stress in this pathosystem, rapid growth rates of the host, and host compensation for disease losses (Charudattan *et al.*, 1985), all of which contribute to lessen the biocontrol efficacy of the pathogen. Since most waterways in this country are intensively managed with governmental tax funds, the cost of inundative strategy is not a significant factor.

5.3.2 *Choice of pathogen: why indigenous; why fungi?*

Pathogens selected to serve as inundative control agents must possess certain essential features. They must be culturable *in vitro* and readily produce spores or other infective propagules that can be used as standardised inoculum. The inoculum must maintain viability (good shelf-life) and be readily capable of infecting (no dormancy factors) and killing the weed in environments with reasonably wide latitude (Daniel *et al.*, 1973). It should be possible to grow the agents in submerged liquid cultures, becuase the current industrial production facilities can meet only this type of fermentation need (Churchill, 1982).

These features are shared by facultative parasites, which can be mass-produced *in vitro*. There are several genera of pathogens among the facultatives that will satisfy both the efficacy and host specificity requirements. However, the fungi have received the most attention for reasons discussed previously (Templeton, 1982; Charudattan, 1985). They offer a wide choice of agents, are usually aggressive parasites, are frequently host-specific, and possess the desired features mentioned above. Most importantly, unlike bacterial and viral pathogens which need insect vectors, natural openings, or wounds for entry into hosts, fungi are capable of active penetration, which makes them the desired candidates for spray applications. Although both indigenous and exotic pathogens may be equally effective as inundative control agents, the choice is clearly with native pathogens (referred to as 'endemic pathogens' by some, which should not be confused with 'endemic diseases'). The native pathogens are exempt from quarantine regulations that govern the use of exotic pathogens for weed control (Charudattan, 1982), and, having co-evolved with native plants and the weed hosts, are likely to be selected for host specificity within the flora and to be adapted for the ecoclimatic conditions of the region. It is therefore easier to find and develop indigenous pathogens as inundative control agents.

5.4 **Status**

Two mycoherbicides, DeVine[R](Abbott) and Collego[TM] (Ecogen), are currently registered and sold in the United States. Two other fungi,

Table 5.1. A list of pathogens under consideration as inundative weed control agents

| Pathogen | Weed[a] | | Location[b] | Reference(s) |
	Common name	Botanical name		
Alternaria crassa	Jimsonweed	*Datura stramonium*	Mississippi, USA	Boyette, 1986
A. helianthi	Cocklebur	*Xanthium strumarium*	Mississippi, USA	P.C. Quimby, personal communication
A. macrospora	Spurred anoda	*Anoda cristata*	Mississippi, USA	Walker & Sciumbato, 1979
A. tenuissima f. sp. *euphorbiae*	Leafy spurge	*Euphorbia esula*	North Dakota, USA	R.M. Hosford, G.D. Statler & L.J. Lutlefield, personal communication
Alternaria sp.	Italian thistle	*Carduus pycnocephalus*	California, USA	S.E. Lindow, personal communication
Ascochyta caulina; *Cercospora chenopodii*	Common lambsquarter	*Chenopodium album*	The Netherlands	Scheepens & van Zon, 1982
Ascochyta pteridis; *Phoma aquilina*, *Cryptomycina pteridis*	Bracken	*Pteridium aquilinum*	California, USA; Scotland, UK	Webb & Lindow, 1981 Irvine *et al.*, 1987
Bipolaris sorghicola	Johnsongrass	*Sorghum halepense*	North Carolina, USA	C.G. Van Dyke, personals communication
Chondrostereum purpureum	Black cherry	*Prunus serotina*	The Netherlands	Scheepens & van Zon, 1982
Collectotrichum coccodes	Velvetleaf	*Abutilon theophrasti*	Vermont, USA;	A.K. Watson & A.R. Gottlieb, personal communication

Pathogen	Common name	Weed[b]	Location[b]	Reference
Colletotrichum dematium	Tall morning glory	*Ipomoea purpurea*	North Carolina, USA	Van Dyke, personal communication
C. dematium f. sp. *crotalariae*; *Fusarium udum* f. sp. *crotalariae*	Showy crotalaria	*Crotalaria spectabilis*	Florida, USA	Charudattan, 1986b
C. gloeosporioides f. sp. *jussiaeae*	Winged waterprimrose	*Jussiaea decurrens*	Arkansas, USA	Boyette *et al.*, 1979
C. malvarum	Prickly sida	*Sida spinosa*	Arkansas, USA	Kirkpatrick, *et al.*, 1982
Colletotrichum sp.	Florida beggarweed	*Desmodium tortuosum*	Georgia, USA	J. Cardina, personal communication.
Colletotrichum sp.	Spiny cocklebur (Bathurst burr)	*Xanthium spinosum*	Orange, Australia	Auld *et al.*, 1986
Curvularia lunata	Barnyardgrass	*Echinochloa crusgalli*	The Netherlands	Scheepens & van Zon, 1982
Fusarium lateritium	Velvetleaf	*Abutilon theophrasti*	Mississippi, USA	Walker, 1981
F. oxysporum f. sp. *cannabis*	Marijuana	*Cannabis sativa*	California, USA	McCain and Noviello 1985
F. solani f. sp. *cucurbitae*	Texas gourd	*Cucurbita texana*	Arkansas, USA	Boyette *et al.*, 1984
Sclerotinia sclerotiorum	Canada thistle	*Cirsium arvense*	Montana, USA	Brosten & Sands, 1986
Septoria silybi	Blessed milkthistle	*Silybum marianum*	California, USA	S.E. Lindow, personal communication

Several other pathogens have been or are being researched worldwide, including many that are suitable candidates for further development as inundative agents; for examples see Charudattan (1978), Scheepens & van Zon (1982), and Templeton (1982).

[a] The common names used in this chapter follow the Composite List of Weeds, *Weed Science*, **32** (Suppl. 2), 1–137, 1984.

[b] Location of the research and the proposed use.

Cercospora rodmanii and *Alternaria cassiae*, have undergone extensive industrial development, and several others, notably *Alternaria macrospora*, *Colletotrichum coccodes*, *C. dematium*, *C. malvarum*, *C. gloeosporioides* f. sp. *jussiaeae, Fusarium lateritium* and *F. solani* f. sp. *cucurbitae*, are being or have been studied from a commercial standpoint in the United States (Table 5.1). Many other fungi are being actively researched in Australia, Canada, Europe and the United States (Table 5.1), confirming the considerable worldwide interest in this approach to weed control (Scheepens & van Zon, 1982; Galbraith & Hayward, 1984; Burge & Irvine, 1985; Auld, 1986; Mortensen, 1986).

5.4.1 *DeVine and Collego*

DeVine is a liquid formulation consisting of chlamydospores of a pathotype of *Phytophthora palmivora* found in the citrus-growing areas of Florida. It is produced and sold by Abbott Laboratories, USA. The research and developmental details of this mycoherbicide have been reviewed previously (Ridings *et al.*, 1976; Charudattan, 1985; Kenney, 1986; Ridings, 1986). It is used as a post-emergent herbicide against *Morrenia odorata* (milkweed vine or stranglervine), an imported weed in Florida citrus groves. It controls seedlings as well as mature vines. Nearly 100% control of the weed is usually obtained, and the control lasts for more than two years. DeVine was registered with the EPA as a mycoherbicide in 1981.

Collego is a wettable powder formulation of *Colletotrichum gloeosporioides* f. sp. *aeschynomene*, and, as stated before, is used for controlling northern jointvetch, *Aeschynomene virginica*, in rice and soybean crops in Arkansas, Mississippi and Louisiana. It is manufactured by the Upjohn Company, and sold since 1988 by Ecogen, USA. The research and developmental details can be found in Daniel *et al.* (1973), Templeton *et al.* (1980, 1984), Churchill (1982), Smith (1982); Charudattan (1985), TeBeest & Templeton (1985) and Smith Jr (1986). Collego is applied aerially or with land-based equipment when the weed has just emerged through the crop canopy. Typical control ranges in excess of 90% (Bowers, 1986), and over the last decade the total area treated with Collego has increased gradually, especially after the banning of the use of 2, 4, 5-T (2-[2, 4, 5-trichlorophenoxy] acetic acid) as a herbicide in rice (Templeton, 1986).

5.4.2 *Cercospora rodmanii*

One of the first mycoherbicide candidates to undergo evaluation and industrial development (by Abbott Laboratories) was *C. rodmanii*, a pathogen of waterhyacinth, *Eichhornia crassipes*, which is the most important aquatic weed in the world. The discovery, research and development of this candidate have been reviewed (Conway *et al.*, 1978; Charudattan, 1984a, 1986a; Freeman & Charudattan, 1984).

Between 1978 and 1982 attempts were made to develop a mycoherbicidal formulation of this fungus with considerable promise. However, the control situation in the United States with respect to waterhyacinth changed dramatically from 1970 to 1980. During this time waterhyacinth became less problematic as a result of the biological control pressure brought on by insects that were introduced in the early 1970s (Charudattan, 1986a). Additionally, the submerged weed hydrilla (*Hydrilla verticillata*) displaced waterhyacinth in most areas where it used to be widespread. Finally, there was continued reliance on chemical herbicides as primary means of control. These factors, coupled with the slowness of the control system based on *Cercospora*, led to the cancellation of attempts to register this fungus as a mycoherbicide. None the less, *C, rodmanii* and the closely related species, *C. piaropi*, are widely distributed in the southeastern United States and are providing considerable biological control of the weed (Freeman & Charudattan, 1974, 1984; Martyn, 1985). However, this natural level of control is not always sufficient to satisfy the public's needs. In view of this, we have proposed the use of *C. rodmanii* as an inundative agent in an integrated weed control scheme involving the insect biocontrols and the application of a limited amount of chemicals (Charudattan, 1986a). If such a scheme were adopted for routine management of waterhyacinth, there would be a need for further industrial development and registration of *C. rodmanii*.

5.4.3 *Alternaria cassiae*

This fungus was first reported from Pakistan in 1960 as a pathogen of *Cassia holosericea*, and was discovered in 1980 as the causal agent of a foliar blight disease of sicklepod (*Cassia obtusifolia*) plants in Mississippi (Walker, 1982). Later it was also found in Florida and other locations (Charudattan *et al.*, 1986) on this weed. In the United States, sicklepod is one of the most important crop weeds, especially in soybean, peanuts and cotton (Charudattan *et al.*, 1986). It is also a troublesome weed in several tropical countries of Africa, the Americas and Asia (Cock & Evans, 1984).

Alternaria cassiae was established as an effective and safe inundative agent against sicklepod by Walker and co-workers (Walker, 1982; Walker & Riley, 1982; Walker & Boyette, 1985). It was further evaluated as a mycoherbicide in a unique region-wide field trial in which five southeastern states participated (Charudattan *et al.*, 1986). The results of this trial, along with Walker's findings (Walker, 1982; Walker & Riley, 1982; Walker & Boyette, 1985), confirmed that *A. cassiae* can control sicklepod seedlings growing in diverse environmental conditions and soil types of the region. The fungus typically yielded 70 to 100% control and performed better than a chemical herbicide, toxaphene (chlorinated camphene). Therefore, *A. cassiae* appears to be an excellent inundative control agent and is being developed as a mycoherbicide by Mycogen Corporation of San Diego, California (Charudattan *et al.*, 1986).

5.4.4 *Other candidates*
Several additional pathogens are being researched as inundative control agents for a variety of weeds in crops, forests and rangelands (Table 5.1). Most of these are in the testing or developmental stage, but it is reasonable to expect that the use of at least some of them will become practical in the near future.

5.5 **Problems**

A number of real and potential problems exist in the practical use of inundative weed control agents. Some of these are related to the biology of the organisms; others are technical, economic or attitudinal problems. Although surmountable through research and education of users, they may none the less delay practice or increase the cost of inundative agents. The following are the important ones and their potential solutions.

5.5.1 *Incompatibility with chemical pesticides*
Deployment of fungal inundative control agents in crops that are intensively managed with chemical pesticides can pose serious problems of incompatibility between the microbial and chemical agents, resulting in a loss of efficacy of the former. In general, mycoherbicides can be expected to be sensitive to fungicides used against crop diseases, but similar incompatibility may also occur with insecticides or chemical herbicides (Anonymous, 1981, 1982; Smith, 1982; Charudattan, 1985; Klerk *et al.*, 1985; Smith Jr, 1986). Some insecticides may be compatible, making possible the tank-mixing of those chemicals and the microbial agent.

Problems of incompatibility have been encountered with Collego in rice and soybean, and with DeVine in citrus (Anonymous, 1981, 1982; Charudattan, 1985). However, they can be overcome by careful integration of pesticides. For example, Smith (1982), Klerk *et al.* (1985) and Smith Jr. (1986) have demonstrated that Collego can be successfully integrated into rice and soybean pest management programmes that involve the use of chemical herbicides to control a spectrum of weeds (other than northern jointvetch), and fungicides and insecticides to manage fungal pathogens and insects affecting the crop. Similarly, DeVine can be integrated into a citrus pest management programme (Anonymous, 1981). In both cases, effective integration depends, with exceptions, on careful sequencing of application of the mycoherbicide before or after certain chemicals. Thus, the fungicides benomyl and triphenyltin hydroxide should be applied to rice one week after applying Collego (Smith Jr, 1986). DeVine must be applied sequentially with wetting agents, fertilisers or pesticides (Anonymous, 1981). On the other hand, Collego can be tank-mixed with the herbicides acifluorfen and bentazon (Smith Jr, 1986); *Cercospora rodmanii* can be tank-mixed with diquat (6, 7-dihydrodipyrido [1, 2-:2', 1'-*c*] pyrazinediium ion) or 2, 4-D ([2, 4-dichlorophenoxy] acetic acid), and *Alternaria cassiae* can be tank-mixed with acifluorfen (R. Charudattan, unpublished). Thus each mycoherbicide and chemical pesticide must be studied in order to develop a set of recommendations for its effective use; the success of the mycoherbicide in the marketplace would depend partly on its ability to be integrated with other pesticides in use.

5.5.2 *Host specificity*
Host specificity of an agent has always been a prime consideration in biological weed control programme. However, the utilisation of facultative parasites as inundative agents raises some important safety questions (Leonard, 1982). For example, will these parasites, which are less host-specialised than obligate parasites, be safe in the long run? How extensive should the host range trials be in order to assess safety? Would these pathogens mutate more readily than obligate parasites? These and similar questions have been discussed by TeBeest & Templeton (1985), Freeman & Charudattan (1985), and Watson (1985), who have expressed cautious optimism.

Significantly, the fungi in Collego and DeVine are not absolutely host specific, i.e. specific only to the target weeds. Since the registration of Collego, several species of *Aeschynomene*, *Pisum sativum*, *Lupinus densiflorus*, *Vicia faba* and *Lathyrus* spp. have been found to be susceptible to *C. gloeosporioides* f. sp. *aeschynomene* (TeBeest & Templeton, 1985). The *Phytophthora palmivora* in DeVine was known during registration to be pathogenic to watermelon, periwinkle and *Rhododendron champanii* in addition to milkweed vine (Anonymous, 1981). Pathogens like these that attack more than one host species nevertheless could be regarded as safe from a host specificity standpoint unlike omnivorous fungi such as *Rhizoctonia solani* and *Sclerotium rolfsii*, provided no hosts of economic, ecological or aesthetic value are attacked by the mycoherbicide candidate. Alternatively, even when absolute host specificity is not available, the agents could still be used in regions where their susceptible non-target hosts are not present. Also, agents that are not readily transported by wind or vectors may be used with little concern for their spread to non-target plants. Finally, the current regulatory thinking in the United States apparently favours 'label restrictions' as a means of warning about potential dangers to non-target plants rather than denying registration for useful mycoherbicidal fungi such as *P. palmivora* (Anonymous, 1981). Thus, the DeVine label restricts the use of this mycoherbicide where susceptible plants are grown and within 100 feet of potentially susceptible plants such as cucumber, squash, begonia, bougainvillea, boxwood, hibiscus, oak, areca palm, *Pittosporum*, snapdragon, *Washingtonia*, coconut palm, and hybrid rhododendron (Anonymous, 1981).

From a marketing perspective, an extreme level of host specificity is generally a disincentive for commercialisation of an agent. Since microbial agents typically control only one out of a spectrum of weeds in a crop, additional herbicides must be used at added expense. A way of overcoming this problem, as mentioned above, is to use a tank mixture of a broad-spectrum chemical herbicide and a mycoherbicide. The primary advantages of tank-mixing are the savings in application costs and improved operational efficiency. But there could be other advantages such as the potential increase in the efficacy of the mycoherbicide by mixing it with chemicals that stress the weed or retard its growth (Charudattan, 1986a; Greaves & Sargent, 1986). Alternatively, as Boyette *et al.* (1979) and Walker (1981) have shown, two pathogens may be combined in one application to control more than one weed, or a pathogen like *Alternaria cassiae*, capable of attacking more than one weed (namely sicklepod, showy crotalaria and coffee senna), can be used to control a group of weeds (Walker, 1983), thereby increasing the market potential of the mycoherbicide agent. Based on their success in experimentally controlling Canada thistle with *Sclerotinia sclerotiorum*, a host-nonspecific fungus with broad host range, Brosten & Sands (1986) have suggested that eventually it may be possible to develop a set of mycoherbicides based on this pathogen through genetic engineering. Although genetic engineering of fungi is more complex than that of pro-karyotes, attempts at modifying host range, virulence, toxin production, etc. in mycoherbicide candidates must begin in earnest if we are to take advantage of the benefits biotechnology has to offer to the field of biological weed control (Charudattan, 1985).

5.5.3 *Resistant weeds*
Theoretically, it should be more difficult to control a genetically hetero-geneous weed with a pathogen than a weed with limited genetic diversity (Burdon & Marshall, 1981; Barrett, 1982). However, because endemic pathogens used as inundative agents are most likely to have co-adapted to the genetic diversity of the host, it should be possible to obtain from nature pathotypes that are virulent for most weed genotypes found in a region. Furthermore, as has been stated (Leonard, 1982; Charudattan, 1985), it should be possible to make strain selections in the laboratory to counter the host resistance encountered in nature.

Experience with Collego, DeVine, *Cercospora rodmanii*, *Alternaria cassiae* and several other mycoherbicide candidates has indicated that host resistance is generally not a significant problem with mycoherbicides. During the past five to fifteen years of use, there has been no emergence of highly resistant or immune genotypes against Collego, DeVine, *A. cassiae* or *C. rodmanii*. Since these pathogens were found among widespread, well established and fairly dominant populations of the respective weed, perhaps they have already been selected for 'excess virulence' against their hosts. In their attempt to select the most destructive pathogen and pathotypes, researchers may have fortuitously selected for the excess virulence. Thus, although resistance to mycoherbicides could be a potential constraint, thus far it has not been of any concern.

5.5.4 *Economic constraints*
Although the cost of developing and registering a mycoherbicide such as Collego is less than that of a chemical herbicide — about $1.5 million to $2.0 million vs. $10 million to $30 million, respectively (Tisdell *et al.*, 1984; Templeton, 1986; R. Charudattan, unpublished) — only private enterprises, not public agencies, can afford to develop inundative control agents (Templeton *et al.*, 1980; Bowers, 1982; Churchill, 1982; Tisdell *et al.*, 1984). However, private industry will necessarily be preoccupied with market size, return on investment and profits. Hence only those inundative agents that show promise of good economic return are likely to be developed.

In this respect, the importance of a patent position, licensing arrangements, industrial collaboration and regulation has been stated (Templeton *et al.*, 1980; Bowers, 1982, 1986; Charudattan, 1982; Churchill, 1982; Saliwanchik, 1986). Patents and licensing protect the invention while providing the industry with exclusive rights to the invention. Yet experience has indicated that the mere patenting of a mycoherbicide candidate is unlikely to induce industrial participation or ultimate commercialisation. Rather, economic realities such as the profitability of the agent, the level of existing and future competition from chemical herbicides, and the technological feasibility of product development will rule the final decision. Therefore, the most promising candidates for future commercialisation are expected to be those capable of solving significant weed problems, not necessarily the major problems, that cannot be solved by chemicals. In other words, weeds for which no chemical herbicides are available or weeds resistant to known chemical herbicides and for which the chemical industry is unwilling to develop a chemical herbicide, will be prime candidates. This would still offer mycoherbicides a niche in the marketplace. Thus Collego on northern jointvetch, because of its small market size and the lack of a suitable chemical herbicide, was more attractive a target for mycoherbicide development. (2, 4, 5-T and silvex (2- [2, 4, 5-trichlorophenoxy] propanoic acid), the chemical herbicides that were in use against northern jointvetch when the Collego agent was discovered, were long under suspicion as being unsafe and were subsequently banned as herbicides, leaving the market open for Collego (Smith Jr, 1986; Templeton, 1986).)

5.5.5 *Efficacy*

Annual crops are characterised by short growth spans during which weed control must be effected. Perennial crops offer greater latitude in this respect, but weed control schedules must still follow a narrow time frame when the weeds are actively growing. Although the inundative control strategy is more suitable for these crops than the inoculative strategy (Charudattan, 1985), it must nevertheless produce rapid and complete or near complete control. In fact, some inundative control agents can act as quickly and effectively as chemical herbicides. For example, *Alternaria cassiae* causes a significant level of seedling mortality within 5 to 10 days after spraying, and 80 to 100% control can be obtained within 3 weeks (Walker, 1982; Walker & Riley, 1982; Charudattan *et al.*, 1986). Both Collego and DeVine require longer periods for control; about 5 weeks for Collego (Anonymous, 1982) and more than 7 weeks for DeVine (Anonymous, 1981).

Cercospora rodmanii on the other hand is slow, needing about 7 months to yield practical levels of control (R. Charudattan, unpublished), and the control is obtained only when host growth rate is slow and if insect biological control agents are also present (Charudattan 1984a, 1986a; Charudattan *et al.*, 1985). Since there are several chemical herbicides available for controlling waterhyacinth and because users are accustomed to fast and nearly complete weed kill soon after chemical treatment, *C. rodmanii* has been expected to perform as well if not better than the existing chemicals. Otherwise, there is no need for the development of an additional agent despite any potential environmental benefits in switching from chemicals to a microbiological agent. The case of *C. rodmanii* therefore exemplifies one of the dilemmas: if the weed could be controlled by an existing chemical herbicide, then the mycoherbicide will be expected to be as efficacious and predictable as the chemical herbicide, the performance of which is generally less affected by environmental variables than that of the biological herbicides.

Although users are becoming more accustomed than in the past to slower-acting herbicides like glyphosate (*N*-[phosphonomethyl] glycine), the length of time it takes for the effectiveness of some of the inundative agents may be unacceptable to users. This puts the mycoherbicide at a competitive disadvantage.

5.6 **Prospects**

From scientific and practical perspectives, inundative control of weeds with indigenous fungi is a successful and promising technology. From the current level of research activities worldwide (Table 5.1) it is reasonable to predict that there will be continued interest in the discovery and development of inundative weed control agents (Scheepens & van Zon, 1982; Templeton, 1982; Charudattan, 1985; TeBeest & Templeton, 1985; Auld, 1986; Mortensen, 1986). More products can be expected to reach the markets, perhaps with *Alternaria cassiae* leading the way. Private industries must, however, continue their commitment and become more actively involved in transferring the technology from laboratory to practical use. A sense of optimism towards the growth of this field has been expressed by many who are actively pursuing research on mycoherbicides, including Quimby & Walker (1982), Scheepens & van Zon (1982), Templeton (1982), Charudattan (1985), and TeBeest & Templeton (1985). It is hoped that this optimism is not unjustified.

Several research priorities need to be addressed in the near future in order to advance this field further (Charudattan, 1985). The most important ones are: (1) integration of chemicals and mycoherbicides; (2) integration of several mycoherbicides as well as mycoherbicides and insect biocontrol agents; (3) genetic improvement of pathotypes for altered host range, increased virulence, toxin production, and resistance/tolerance to chemical pesticides; and (4) development and registration of additional mycoherbicides.

References

Anonymous (1981). *DeVine Label*. Chemical & Agricultural Products Division, Abbott Laboratory, Chicago, IL, 1 p.
Anonymous (1982). *Collego Technical Manual*. Tuco, The Upjohn Company, Agricultural Division, Kalamazoo, MI, 24 pp.
Auld, B.A. (1986). *Potential for Mycoherbicides in Australia*. Workshop Proceedings, Agricultural Research and Veterinary Centre, Orange, NSW, Australia, 44 pp.
Auld, B.A., McRae, C.F. & Nikandrow, A. (1986). Current research on pathogens as mycoherbicides for *Xanthium* spp. In: *Potential for Mycoherbicides in Australia*, ed. B.A. Auld, pp. 20–1. Agricultural Research and Veterinary Centre, Orange, NSW, Australia.
Barrett, S.C.H. (1982). Genetic variation in weeds. In: *Biological Control of Weeds with Plant Pathogens*, eds R. Charudattan & H.L. Walker, pp. 73–98. Wiley, New York.
Bemmann, W. (1985). From weeds isolated fungi and their capability for weed control — a review of publications. *Zentralblatt für Mikrobiologie*, **140**, 111–48.
Bowers, R.C. (1982). Commercialization of microbial biological control agents. In:

Biological Control of Weeds with Plant Pathogens, eds R. Charudattan & H.L. Walker, pp. 157–73. Wiley, New York.

Bowers, R.C. (1986). Commercialization of Collego — an industrialist's view. *Weed Science*, **34** (Suppl. 1), 24–5.

Boyette, C.D. (1986). Evaluation of *Alternaria crassa* for biological control of jimsonweed: host range and virulence. *Plant Science*, **45**, 223–8.

Boyette, C.D., Templeton, G.E. & Oliver, L.R. (1984). Texas gourd (*Cucurbita texana*) control with *Fusarium solani* f. sp. *cucurbitae*. *Weed Science*, **32**, 649–55.

Boyette, C.D., Templeton, G.E. & Smith, R.J. (1979). Control of winged waterprimrose (*Jussiaea decurrens*) and northern jointvetch (*Aeschynomene virginica*) with fungal pathogens. *Weed Science*, **27**, 497–501.

Brosten, B.S. & Sands, D.C. (1986). Field trials of *Sclerotinia sclerotiorum* to control Canada thistle (*Cirsium arvense*). *Weed Science*, **34**, 377–80.

Bruckart, W.L. & Dowler, W.M. (1986). Evaluation of exotic rust fungi in the United States for classical biological control of weeds. *Weed Science*, **34** (Suppl. 1), 11–14.

Burdon, J.J. (1982). The effect of fungal pathogens on plant communities. In: *The Plant Community as a Working Mechanism*, ed. E.I. Newmann, pp. 99–112. Blackwell Scientific Publications, Oxford.

Burdon, J.J. & Marshall, D.R. (1981). Biological control and the reproductive mode of weeds. *Journal of Applied Ecology*, **18**, 649–58.

Burge, M.N. & Irvine, J.A. (1985). Recent studies on the potential for biological control of bracken using fungi. *Proceedings of the Royal Society of Edinburgh*, **86B**, 187–94.

Charudattan, R. (1978). *Biological Control Projects in Plant Pathology — A Directory*. Misc. Publ., Plant Pathol. Dept, Univ. Florida, Gainesville, FL, 67 pp.

Charudattan, R. (1982). Regulation of microbial weed control agents. In: *Biological Control of Weeds with Plant Pathogens*, eds. R. Charudattan & H.L. Walker, pp. 175–88. Wiley, New York.

Charudattan, R. (1984a). Role of *Cercospora rodmanii* and other pathogens in the biological and integrated controls of water hyacinth. In: *Proceedings of the International Conference on Water Hyacinth*, ed. G. Thyagarajan, pp. 834–59. United Nations Environment Programme, Nairobi, Kenya.

Charudattan, R. (1984b). Microbial control of plant pathogens and weeds. *Journal of Georgia Entomological Society*, **19** (Suppl. 2), 40–62.

Charudattan, R. (1985). The use of natural and genetically altered strains of pathogens for weed control. In: *Biological Control in Agricultural IPM Systems*, eds M.A. Hoy & D.C. Herzog, pp. 347–72. Academic Press, Orlando, FL.

Charudattan, R. (1986a). Integrated control of waterhyacinth with a pathogen, insects, and herbicides. *Weed Science*, **34**, (Suppl. 1), 26–30.

Charudattan, R. (1986b). Biological control of showy crotalaria (*Crotalaria spectabilis*) with two fungal pathogens. *WSSA Abstracts*, **26**, 51.

Charudattan, R., Linda, S.B., Kluepfel, M. & Osman, Y.A. (1985). Biocontrol efficacy of *Cercospora rodmanii* on waterhyacinth. *Phytopathology*, **75**, 1263–9.

Charudattan, R. & Walker, H.L. (1982). *Biological Control of Weeds with Plant Pathogens*. Wiley, New York, 293 pp.

Charudattan, R., Walker, H.L., Boyette, C.D., Ridings, W.H., TeBeest, D.O., Van Dyke, C.G. & Worsham, A.D. (1986). Evaluation of *Alternaria cassiae* as a mycoherbicide for sicklepod (*Cassia obtusifolia*) in regional field tests. *Southern*

Cooperative Series Bulletin 317, Alabama Agric. Exp. Stn, Auburn Univ., AL, 19 pp.

Charudattan, R., Zettler, F.W., Cordo, H.A. & Christie, R.G. (1980). Partial characterization of a potyvirus infecting the milkweed vine, *Morrenia odorata*. Phytopathology, **70**, 909–13.

Churchill, B.W. (1982). Mass production of microorganisms for biological control. In: *Biological Control of Weeds with Plant Pathogens*, eds R. Charudattan & H.L. Walker, pp. 139–56. Wiley, New York.

Cock, M.J.W. & Evans, H.C. (1984). Possibilities for biological control of *Cassia tora* and *C. obtusifolia*. *Tropical Pest Management*, **30**, 339–50.

Conway, K.E., Cullen, R.E., Freeman, T.E. & Cornell, J.A. (1979). Field evaluation of *Cercospora rodmanii* as a biological control of waterhyacinth. Misc. Paper A-79-6, US Army Waterways Experiment Station, Vicksburg, MS, 46 pp.

Conway, K.E., Freeman, T.E. & Charudattan, R. (1978). Development of *Cercospra rodmanii* as a biological control for *Eichhornia crassipes*. In: *Proceedings EWRS 5th Symposium on Aquatic Weeds*, ed. Anonymous, pp. 225–30. EWRS Secretariat, P.O. Box 14, Wageningen, The Netherlands.

Cullen, J.M., Kable, P.F. & Katt, M. (1973). Epidemic spread of a rust imported for biological control. *Nature*, **244**, 462–4.

Daniel, J.T., Templeton, G.E., Smith, R.J. & Fox, W.T. (1973). Biological control of northern jointvetch in rice with an endemic fungal disease. *Weed Science*, **21**, 303–7.

Dyer, W.E., Turner, S.K., Fay, P.K., Sharp, E.L. & Sands, D.C. (1982). Control of Canada thistle by a rust, *Puccinia obtegens*. In: *Biological Control of Weeds with Plant Pathogens*, eds R. Charudattan & H.L. Walker, pp. 243–4. Wiley, New York.

Emge, R.G. & Templeton, G.E. (1981). Biological control of weeds with plant pathogens. In: *Biological Control in Crop Production*, ed. G. Papavizas, pp. 219–26. Allanheld, Osmun & Co. Totowa, NJ.

Freeman, T.E. & Charudattan, R. (1974). Occurrence of *Cercospora piaropi* on waterhyacinth in Florida. *Plant Disease Reporter*, **58**, 277–78.

Freeman, T.E. & Charudattan, R. (1984). *Cercospora rodmanii* Conway: a biocontrol agent for waterhyacinth. *Technical Bulletin 842*, Agric. Exp. Stn, Univ. Florida, Gainesville, FL, 18 pp.

Freeman, T.E. & Charudattan, R. (1985). Conflicts in the use of plant pathogens as biocontrol agents for weeds. In: *Proceedings of the VIth International Symposium on Biological Control of Weeds*, Vancouver, Canada, 19–25 August 1984, Vancouver, Canada, ed. E.S. Delfosse, pp. 351–7. Agriculture Canada, Canadian Government Publishing Centre, Ottawa, Canada.

Galbraith, J.C. & Hayward, A.C. (1984). *The Potential of Indigenous Micro-organisms in the Biological Control of Water Hyacinth in Australia*. A Project Report, Department of Resources and Energy, Canberra, Australia, 84 p.

Greaves, M.P. & Sargent, J.A. (1986). Herbicide-induced microbial invasion of plant roots. *Weed Science*, **34** (Suppl. 1), 50–3.

Hasan, S. (1980). Plant pathogens and biological control of weeds. *Review of Plant Pathology*, **59**, 349–56.

Hasan, S. & Wapshere, A.J. (1973). The biology of *Puccinia chondrillina*, a potential biological control agent of skeleton weed. *Annals of Applied Biology*, **74**, 325–32.

Holcomb, G.E. (1982). Constraints on disease development. In: *Biological Control of Weeds with Plant Pathogens*, eds R. Charudattan & H.L. Walker, pp. 61–71. Wiley, New York.

Irvine, J.I.M., Burge, M.N. & McElwee, Marion (1987). Association of *Phoma aquilina* and *Asochyte pteridis* with curl-tip disease of bracken. *Annals of Applied Biology*, **110**, 25–31.

Kenney, D.S. (1986). DeVine — the way it was developed — an industrialist's view. *Weed Science*, **34** (Suppl. 1), 15–16.

Kirkpatrick, T.L., Templeton, G.E., TeBeest, D.O. & Smith, R.J. (1982). Potential of *Colletotrichum malvarum* for biological control of prickly sida. *Plant Disease*, **66**, 323–5.

Klerk, R.A, Smith, R.J. & TeBeest, D.O. (1985). Integration of a microbial herbicide into weed and pest control programs in rice. *Weed Science*, **33**, 95–9.

Leonard, K.J. (1982). The benefits and potential hazards of genetic heterogeneity in plant pathogens. In: *Biological Control of Weeds with Plant Pathogens*, eds R. Charudattan & H.L. Walker, pp. 99–112. Wiley, New York.

McCain, A.H. & Noviello, C. (1985). Biological control of *Cannabis sativa*. In: *Proceedings of the VIth International Symposium on Biological Control of Weeds*, 19–25 August 1984, ed. E.S. Delfosse, pp. 635–42. Agriculture Canada, Vancouver.

Martyn, R.D. (1985). Waterhyacinth decline in Texas caused by *Cercospora piaropi*. *Journal of Aquatic Plant Management*, **23**, 29–32.

Massion, C.L. & Lindow, S.E. (1986). Effects of *Sphacelotheca holci* infection on morphology and competitiveness of johnsongrass (*Sorghum halepense*). *Weed Science*, **34**, 883–8.

Mortensen, K. (1986). Biological control of weeds with plant pathogens. *Canadian Journal of Plant Pathology*, **8**, 229–31.

Oehrens, E. (1977). Biological control of the blackberry through the introduction of rust, *Phragmidium violaceum*, in Chile. *FAO Plant Protection Bulletin*, **25**, 26–8.

Oehrens, E. & Gonzales, S. (1974). Introduction of *Phragmidium violaceum* (Schulz) Winter as a biological control agent for zarzamora (*Rubus constrictus* Lef. and M. and *R. ulmifolius* Schott.) (in Spanish). *Agro Sur (Chile)*, **2**, 30–3.

Olivieri, I. (1984). Effect of *Puccinia cadui-pycnocephali* on slender thistles (*Carduus pycnocephalus* and *C. tenuiflorus*). *Weed Science*, **32**, 508–10.

Phatak, S.C., Sumner, D.R., Wells, H.D., Bell, D.K. & Glaze, N.C. (1983). Biological control of yellow nutsedge with the indigenous rust fungus *Puccinia canaliculata*. *Science*, **219**, 1446–7.

Politis, D.J., Watson, A.K. & Bruckart, W.L. (1984). Susceptibility of musk thistle and related composites to *Puccinia carduorum*. *Phytopathology*, **74**, 687–91.

Quimby, P.C. & Walker, H.L. (1982). Pathogens as mechanisms for integrated weed management. *Weed Science*, **30** (Suppl. 1) 30–4.

Ridings, W.H. (1986). Biological control of stranglervine in citrus — a researcher's view. *Weed Science*, **34** (Suppl. 1), 31–2.

Ridings, W.H., Mitchell, D.J., Schoulties, C.L. & El-Gholl, N.E. (1976). Biological control of milkweed vine in Florida citrus groves with a pathotype of *Phytophthora citrophthora*. In: *Proceedings of the IVth International Symposium on Biological Control of Weeds*, ed. T.E. Freeman, pp. 224–40. Univ. Florida, Gainesville, FL.

Saliwanchik, R. (1986). Patenting/licensing of microbiological herbicides. *Weed Science*, **34** (Suppl. 1), 43–9.

Scheepens, P.C. & van Zon, H.C.J. (1982). Microbial herbicides. In: *Microbial and Viral Pesticides*, ed. E. Kurstak, pp. 623–41. Marcel Dekker, New York.

Shrum, R.D. (1982). Creating epiphytotics. In: *Biological Control of Weeds with Plant Pathogens*, eds R. Charudattan & H.L. Walker, pp. 113–36. Wiley, New York.

Smith, R.J. (1982). Integration of microbial herbicides with existing pest management programs. In: *Biological Control of Weeds with Plant Pathogens*, eds R. Charudattan & H.L. Walker, pp. 189–203. Wiley, New York.

Smith, R.J. Jr (1986). Biological control of northern jointvetch in rice and soybeans — a researcher's view, *Weed Science*, **34**, (Suppl. 1), 17–23.

TeBeest, D.O. & Templeton, G.E. (1985). Mycoherbicides: progress in the biological control of weeds. *Plant Disease*, **69**, 6–10.

Templeton, G.E. (1982). Status of weed control with plant pathogens. In: *Biological Control of Weeds with Plant Pathogens*, eds R. Charudattan and H.L. Walker, pp. 29–44. Wiley, New York.

Templeton, G.E. (1986). Mycoherbicide research at the University of Arkansas — past, present and future. *Weed Science*, **34** (Suppl. 1), 35–7.

Templeton, G.E. & Greaves, M.P. (1984). Biological control of weeds with fungal pathogens. *Tropical Pest Management*, **30**, 333–8.

Templeton, G.E., Smith, R.J. & Klomparens, W. (1980). Commercialization of fungi and bacteria for biological control, *Biocontrol News and Information*, **1**, 291–4.

Templeton, G.E., TeBeest, D.O. & Smith, R.J. Jr (1979). Biological weed control with mycoherbicides. *Annual Review of Phytopathology*, **17**, 301–10.

Templeton, G.E., TeBeest, D.O. & Smith, R.J., Jr (1984). Biological weed control in rice with a strain of *Colletotrichum gloeosporioides* (Penz.) Sacc. used as a mycoherbicide. *Crop Protection*, **3**, 409–22.

Templeton, G.E., Smith, R.J., Jr & TeBeest, D.O. (1987). Progress and potential of weed control with mycoherbicides. *Reviews in Weed Science*, **2**, 1–14.

Tisdell, C.A., Auld, B.A. & Menz, K.M. (1984). On assessing the value of biological control of weeds. *Protection Ecology*, **6**, 169–79.

Trujillo, E.E. (1985). Biological control of hamakua pamakani with *Cercosporella* sp. in Hawaii. In: *Proceedings of the VIth International Symposium on Biological Control of Weeds*, 19–25 August, 1984, ed. E.S. Delfosse, pp. 661–71. Agriculture Canada, Vancouvers.

Van der Plank, J.E. (1963). *Plant Diseases: Epidemics and Control*. Academic Press, New York.

Van der Plank, J.E. (1975). *Principles of Plant Infection*. Academic Press, New York.

Walker, H.L. (1981). *Fusarium lateritium*: a pathogen of spurred anoda (*Anoda cristata*), prickly sida (*Sida spinosa*), and velvetleaf (*Abutilon theophrasti*), *Weed Science*, **29**, 629–31.

Walker, H.L. (1982). A seedling blight of sicklepod caused by *Alternaria cassiae*. *Plant Disease*, **66**, 426–8.

Walker, H.L. (1983). Control of sicklepod, showy crotalaria, and coffee senna with a fungal pathogen. US Patent No. 4,390,360.

Walker, H.L. & Boyette, C.D. (1985). Biological control of sicklepod (*Cassia obtusifolia*) in soybeans (*Glycine max*) with *Alternaria cassiae*. *Weed Science*, **33**, 212.

Walker, H.L. & Riley, J.A. (1982). Evaluation of *Alternaria cassiae* for the biocontrol of sicklepod (*Cassia obtusifolia*). *Weed Science*, **30**, 651–4.

Walker, H.L. & Sciumbato, G.L. (1979). Evaluation of *Alternaria macrospora* as a potential biological control agent for spurred anoda (*Anoda cristata*): host range studies. *Weed Science*, **27**, 612–14.

Walker, J.C. (1969). *Plant Pathology*, 3rd edn. McGraw-Hill, New York, 819 pp.

Watson, A.K. (1985). Host specificity of plant pathogens in biological weed control. In: *Proceedings of the VIth International Symposium on the Biological Control of Weeds*, 19–25 August 1984, ed. E.S. Delfosse, pp. 577–86. Agriculture Canada, Vancouver.

Watson, A.K. & Clement, M. (1986). Evaluation of rust fungi as biological control agents of weedy *Centaurea* in North America. *Weed Science*, **34** (Suppl. 1), 7–10.

Webb, R.R. and Lindow, S.E. (1981). Evaluation of *Assochyta pteridis* as a potential biological control agent of bracken fern (Abstr.). *Phytopathology* **71**, 911.

Fungi in classical biocontrol of weeds

6.1 Introduction

The intentional use of plant pathogens and, more specifically, of plant pathogenic fungi to control weeds is a relatively recent event in plant pathology and weed science (Freeman *et al.*, 1978; Templeton, 1982). It is an attempt to integrate these two sciences to bring about control of specific weeds using disease-causing organisms. In this integration of weed science and plant pathology, two distinct tactics are emerging for the use of plant pathogens to control weeds: the classical or traditional biological control tactic, and the inundative or mycoherbicide tactic (Templeton & Trujillo, 1981). Classical biological control of weeds using fungi is the intentional introduction of one or more organisms which attack the weed in its native range, or elsewhere, to a region where the weed exists at noxious levels in the absence of the organisms (Wapshere, 1979).

The inundative tactic uses massive inoculations of fungi which are usually indigenous to the area. These pathogenic fungi are propagated artificially on a large scale and then applied in a manner similar to herbicides, hence the term mycoherbicide. Each plant is inundated with massive doses of inoculum at the optimum time for infection, thus creating an epidemic with a pathogen that normally persists at an endemic level. The inundative tactic is discussed by Charudattan (Chapter 5).

Control tactics which have been developed to cope with weeds in cultivated fields include mechanical, cultural, chemical and mycoherbicidal methods. In rangeland or non-cropland it is often not economical to use any of these control methods since the costs associated with such control may far exceed the benefit received or even the actual value of these lands. Classical biological control, which depends on the fungus spreading itself throughout the population of weeds, is a 'low cost' alternative for the land owner. While cost effective for the land owner, there is little profit motive for private industry to develop a biological control agent for a 'once only' application, with the result that this work must be publicly supported.

In addition to economic considerations, the increasing awareness of the potential and realised hazards associated with herbicide use have made alternatives and supplements to chemical weed control practices more desirable. Wilson (1969) lists three advantages that the use of plant pathogens would have over herbicides: (1) they can be specific to the weeds, (2) residue and toxicity problems would be reduced or eliminated altogether, and (3) there would be no accumulation of the herbicides in the soil or groundwater.

Classical biological control has been used with great success in the past by entomologists who have released many insects into weed populations where no pests have previously existed. Although interest in, and use of, plant pathogenic fungi to control weeds has increased in the past 20 years, its practice is lagging far behind the use of insects. This appears to reflect a lack of research effort rather than a lack of suitable fungi. Although public interest in classical biological control is increasing, particularly in rangeland areas, there are few extensive integrated weed control programmes to combat the spread of alien plant species.

6.2 Selection of fungi for classical biological control

Plant pathogenic fungi account for only 5% of all of the biological control agents being used or being considered for use in controlling weeds. Of these fungi, less than one-half are exotic fungi for classical biological control projects; the rest come into the category of mycoherbicides. Pathogenic fungi for classical biological control are sought from the original geographical range of the weed and screened for virulence, and promising candidate organisms are introduced into the new region with the expectation that they will become established and increase to epiphytotic levels. Ideally a successful pathogen would become endemic to the region and continuously suppress the introduced weed to subeconomic levels. Pathogenic fungi may also be sought from host plant species in other parts of the world which are related to native weed species that have evolved in the absence of such pathogens. Thus, the native weed species would have little or no resistance to the introduced fungi. The rationale for these new parasite–host associations comes from the historical experience of diseases such as chestnut blight, white pine blister rust, coffee rust and potato late blight, which were caused by accidental introduction of fungal pathogens on to desirable plants (Wilson, 1969). Hokkanen & Pimentel (1984) found that the success rate resulting from the introduction of natural enemies, specifically insects, for biological control was 75% higher using this method of employing new parasite–host associations. However, Goeden & Kok (1986) argue that this approach is not effective for selecting control agents and should not be used as the preferred method. This is borne out by the fact that new fungus–host associations have not yet produced significant results in weed control projects.

Candidates for classical biological control projects are usually sought in the original or evolutionary home of the weed. Table 6.1 lists a number of projects currently underway or being considered. With few exceptions the fungi imported to control weed pests are introduced from the origin of the weed. The exceptions appear to be collections of pathogens from inadvertent introductions of the pathogen along with the weed into a new region. The candidate fungal pathogen must be highly effective and only affect the weed species intended for control. When *Puccinia chondrillina* Bubak. & Sydow. was imported into Australia, 62 species were tested for susceptibility to the rust before it was considered that the fungus was specific to *Chondrilla juncea* L. At that time, however, concern was expressed regarding the adequacy of the tests, and consequently protocols for testing and quarantine of new biological control agents were established by most countries involved in these projects.

Wapshere (1979) has outlined a strategy for demonstrating the safety of new biological control agents for weeds. The new agent should be tested against a group of phylogenetically related plants plus those socially important plants which may be considered at risk for one of several reasons including the fact that (1) the plants are related to the weed, (2) the mycology related to the new agent and the 'at risk' plant is inadequate, (3) the plant is attacked by a close relative of the new agent, or (4) the plant has never been ecologically exposed to the new agent. This method of testing for safety is replacing the older method of exposing a broad spectrum of crop plants to the new agent. The new method considers the safety of not only crop plants but also other desirable plants whether they are introduced or native species.

Caution is well warranted. For example, there are two rusts of diffuse knapweed, *Centauria diffusa* Lam., which would be highly desirable as biological agents to help control this rapidly spreading weed in North

Table 6.1. Some examples of fungi used or being considered for use in biological control projects

Weed species	Origin	Fungus	Origin	Released	Country
Ageratina riparia (hamakua pamakani)	Mexico	*Cercosporella ageratinae*	Jamaica	1975	USA (Hawaii)
Ambrosia trifida (giant ragweed)	North America	*Puccinia xanthii*	North America	—[a]	Europe
Carduus nutans (musk thistle)	Eurasia	*Puccinia carduorum*	Eurasia	—[a]	USA
C. pycnocephalis (Italian thistle)	Eurasia	*Puccinia cardui-pycnocephali*	France	—[a]	USA
C. tenuiflorus (slenderflower thistle)	Eurasia	*Puccinia carduorum*	Eurasia	—[a]	USA
Chondrilla juncea (rush skeletonweed)	Europe	*Puccinia chondrillina*	Europe	1971 1978	Australia, USA
Cirsium arvense (Canada thistle)	Europe	*Puccinia xanthii*	North America	prior to 1974[b]	Australia
	Europe	*Puccinia punctiformis*	Europe	prior to 1915[b]	New Zealand

Weed		Rust fungus			
Cyperus esculentus (yellow nutsedge)	Eurasia	*Puccinia canaliculata*	Europe	—[a]	USA (Florida)
Eichhornia crassipes (water-hyacinth)		*Uredo eichhorniae*	Argentina	—[a]	USA
Euphorbia cyparissias (cypress spurge)	Eurasia	*Uromyces alpestris*	Europe	—[a]	USA
		Uromyces scutellatus	Europe	—[a]	USA
Euphorbia esula (leafy spurge)	Eurasia	*Uromyces striatus*	Europe	—[b]	USA
Galega officinalis (goatsrue)	Asia Minor	*Uromyces galega*	France	1973	Chile
Rubus constrictus (blackberry)	Asia, Europe	*Phragmidium violaceum*	Germany	1973	Chile
R. fruticosus	Europe	*Phragmidium violaceum*	Germany	1983	Australia
R. ulmifolius	Asia, Europe	*Phragmidium violaceum*	Germany	1973	Chile

[a] Under study but not released.
[b] Accidental introduction.

America. Watson *et al.* (1981) have demonstrated that diffuse knapweed is susceptible to *Puccinia centaurea* DC. and *P. jaceae* Otth. (which are possibly synonyms of the same organism). However, safflower, which is of course a desirable crop, was also shown to be susceptible to *Puccinia jaceae*. Leafy spurge is susceptible to *Uromyces striatus* Schroet., which kills nearly all infected plants. This rust occurs rarely in nature and its alternate host is alfalfa (Littlefield, 1985). Consequently these rusts will probably never be released.

There is, none the less, a rising controversy in the scientific community over the validity of the screening tests which have elminated the above mentioned fungi and others as biological control agents. When the fungi are screened for virulence on other plant species, the testing occurs under artifically controlled environmental conditions. Often these fungi exhibit a wider host range under artificial conditions than has been observed in nature (Watson, 1985). For example, under artificial conditions, *P. junceae* infects the cotyledons and lower leaves of safflower (*Carthamnus tinctorius* L.), producing viable pustules on the cotyledons. Resistant reactions are observed on a few lower leaves and no infectons develop on the upper portions of the plant. Under field conditions, however, the rust has never been observed (Mortensen, 1985). Another example is the susceptibility of musk thistle to *P. carduorum*, which under controlled conditions also infects artichoke (*Cynara scolymus* L.). Mature plant resistance has been observed in artichoke in the field, and the rust has not been reported on this plant in Eurasia. Susceptibility therefore appears to be an artefact of glasshouse conditions, and the rust probably poses little risk to artichoke production in North America (Bruckart *et al.*, 1985). It could thus be argued that the potential benefits of biological control using such agents are not being realised because of the unrealistic findings on host range obtained under controlled environmental conditions. This author would encourage continued caution, however, because one error resulting in crop damage could jeopardise the future of all biological control programmes.

Public awareness is also influencing decisions on the potential release of new biological control agents. For example, the European blackberry is being considered for a biological control project in Australia, and a controversy has developed concerning possible introductions of the rust fungus *Phragmidium violaceum* C.F. Schultz. as a control agent. Losses to ranchers resulting from infestations by the blackberry exceed $30 million annually, but on the other hand the plant has brought benefits approaching $1 million to berry canners and bee keepers (Field, 1985). However, it appears that the quarantine laws of Australia have been circumvented since outbreaks of the rust have been reported in localised, heavy infestations which have left little doubt that they were intentionally released (Bruzzese, 1984).

Another factor in the selection of a suitable fungal agent for classical biological control is the mobility of the organism, which must be sufficient to allow its rapid spread through the target weed population. In the most suitable areas for classical biological control, i.e. pastures and rangeland, weed populations are not as evenly distributed as they might be in cultivated situations. Weed populations under these conditions tend to be grouped in favourable sites and may be separated by distances of a few metres to a few kilometres. Consequently dispersal propagules must be able to travel large distances from host plant to host plant, particularly at the edges of expanding infestations. Large-scale movement may also have to be aided by man when spread of the fungus is limited by natural barriers. Propagules which are adapted for wind dispersal are the most suitable, whereas those dependent on other dispersal mechanisms such as splashing raindrops may not be able to spread as rapidly throughout the weed population. This is one reason why fungi, particularly the rusts, have been the main thrust of research in this field. This is evident from Table 6.1 in which all but one of the exotic fungi are rusts. The rusts are particularly suitable as biological control agents for several additional reasons: they are generally very host specific; they are not likely to attack either native or crop plants; and they are capable of producing large quantities of spores which are generally wind disseminated. Very few other types of plant pathogen have shown the same potential as the rusts.

The degree of success of an exotic fungus imported into a new region depends on its ability to adapt to the climate of the new region, which will influence various factors in the disease cycle. When *P. chondrillina* was released on rush skeletonweed in Australia, it spread 320 km in the first year (Cullen *et al.*, 1973), whereas in the state of Washington, USA, the rust had spread less than 10 km in its first three years (Adams & Line, 1984a). In Australia the warm and relatively wet climate of summer aids the dispersal of the rust. Washington, however, receives 70% of its precipitation during the winter months with the result that relatively high springtime populations of the rust decline rapidly as the weather becomes dryer and the plants bolt for flowering. Rush skeletonweed is highly adapted for dry summer climates; most of the leaves are shed as the climate becomes dryer and the inoculum, as well as the diseased leaves, are lost so that disease is minimal. Rust populations may increase in the autumn but the time intervals between generations of spores and subsequent spread are very slow under cool, autumn conditions. Consequently, weed populations have not declined as rapidly as in Australia.

In order that successful control can be achieved, it is also important that the ability of the fungus to spread is greater than that of the weed. If the fungal pathogen remains localised while the weed continues to spread, little is to be gained. Additional spread of the pathogen to new weed infestations must then be accomplished by man.

The fungal pathogen should ideally be able to kill the target weed species, but more often an additional selection pressure is necessary. Even though the fungus may not kill the target weed, it may slow its rate of spread, thereby conserving some time in which other control measures may be brought to bear.

6.3 Control of specific weeds

6.3.1 *Established programmes*
The first intentional use of a plant pathogenic fungus to control a weed occurred in Australia when Hasan (1970) suggested the possibility of introducing *P. chondrillina* into that country for control of rush skeleton-weed. After much research and deliberation, the rust was finally released through quarantine in the summer of 1971–72 at Wagga Wagga (Cullen *et al.*, 1973).

The rust appears to exist as several races which are very specific to different types of the weed which are dissimilar either phenotypically or in terms of resistance or a combination of both. *Chondrilla juncea* is an apomictic plant and populations of the weed in a region are in reality clones of the original introduction. Therefore, the weed population may be either susceptible or resistant to a particular race of the introduced rust. In Australia, the rust was quite specific to the predominant type (type 'A') of the weed, with the result that the fungus spread rapidly through the country infecting only this type. Cullen (1976) reported that the rust disease was not highly damaging to the rush skeletonweed population during the first four years following its introduction, although there was a decline in the mean number of plants per 2 square metre at four different sites from 170 to 50.

Types 'B' and 'C' of skeletonweed were initially of minor importance in the area infested and remained completely free of rust. However, as the population of type 'A' became effectively reduced by the rust, these uninfected types have tended to increase in number and invade the eco-logical niche previously occupied by type 'A' (Schmidl, 1980). Cullen (1985) reported a continuation of this trend, and the lack of a satisfactory biological control for types 'A' and 'B' has adversely affected the predicted returns of the overall programme in Australia.

In the United States, the first use of an exotic plant pathogen for weed control was also that of *P. chondrillina* to control rush skeletonweed (Emge *et al.*, 1981). Three, possibly four, distinct forms of the weed are found in the western United States showing different degrees of suscepti-bility to the fungus. Virulent isolates of the rust were selected, increased, and supplied to workers in the western United States for release on *C. juncea* in specific areas. Inoculations resulted in rust initiation, pathogen survival over winter and/or summer in each area, and natural spread of the pathogen to uninoculated areas. In Washington and Idaho, although population declines have not yet been documented, the disease is having an effect: there is a reduction in plant height, number of flowers and seed viability. Flower-bud formation is reduced by approximately 50% and seed viability is decreased by 40% when pustules cover only 5% of the plant's surface (Adams & Line, 1984b). Rosettes that become heavily infected may die before flowering, and seedling mortality may reach 90% in areas where the climate is conducive to disease development (Lee, 1986).

Another weed which has been successfully controlled by an introduced fungal pathogen is the blackberry. The rust *Phragmidium violaceum* (Schultz) Winter was released in Chile in 1973 on two species of blackberry, *Rubus constrictus* Lefevre et P.J. Mueller, and *R. ulmifolius* Schott. (Oehrens & Gonzales, 1974, 1977; Oehrens, 1977). The rust has disseminated rapidly and has reduced weed populations, especially in the case of *R. constrictus* which is more susceptible than *R. ulmifolius*. The rust causes defoliation and prevents proper lignification of stems, which leads to increased frost damage and invasion by secondary pathogens.

A further example of success with a rust (also in Chile) is the use of *Uromyces galega* (Opiz) Saccardo, which is reported to be established and spreading on goatsrue, *Galega officinales* L. (Oehrens & Gonzales, 1975).

Most achievements in classical biological control with fungi involve the use of rust pathogens. However, a few successful programmes have involved Fungi Imperfecti. Nakao & Funasaki (1979), for example, recorded the release of *Cercosporella ageratinae* (nomen nudem) in 1975 for the control of hamakua pamakani (*Ageratina riparia* (Regal) R.M. King & H. Robinson). The release on the islands of Oahu, Maui and Hawaii in the state of Hawaii, USA, has resulted in a 40 to 60% reduction in weed populations and the decline is continuing. Also in Hawaii, Trujillo & Obrero (1972, 1978) have demonstrated the use of a species of *Cephalosporium* to control kolomona brush weed (*Cassia surattensis*) in pastures. The fungus causes a vascular wilt, and, if direct inoculations were employed, large 4- to 5-year-old bushes died in 3–4 months. Natural spread occurred more slowly.

6.3.2 *Promising new strategies*

The examples mentioned so far are all of established and successful classical weed control strategies with fungi. Several others with obvious potential are currently under investigation. For example, thistles are particularly aggressive and a number of species of *Carduus* are currently being studied as candidates for biological control using *Puccinia* spp. in Australia. These include the slenderflower thistle *C. tenuiflorus* Curt., the Italian thistle *C. pycnocephalus*, and musk thistle *C. nutans* L. Slenderflower and Italian thistles are susceptible to a single rust species, *P. cardui-*

pycnocephali Sydow, which is known only on the genus *Carduus*. Growth of the Italian thistle, but not of the slenderflower thistle, was affected in trials (Olivieri, 1984). The Australian forms of Italian thistle were more susceptible than the specimens from France where the original rust samples were collected. Thus, *P. cardui-pycnocephali* may be a good candidate for importation into Australia.

Another rust, *P. carduorum* Jacky, attacks the musk thistle (Politis *et al.*, 1984) and in California there is a race of this rust which appears to be host specific to the slenderflower thistle (Watson & Brunetti, 1985). It appears that this rust was inadvertently introduced prior to recent efforts to find a biological control agent.

A biological control for *Cirsium arvense* L. (Canada or creeping thistle) has been sought for many years. Rust diseases of this aggressive weed have been proposed for its control since 1894 (Halstead, 1894). *Puccinia obtegens* (Link.) Tul., for example, has been observed to kill individual plants or small stands (Turner *et al.*, 1982) and *P. punctiformis* Strauss (Roehl.), which was accidentally introduced into New Zealand prior to 1915, retards growth and reduces the vigour of the weed when the spread of spores is assisted (Julien, 1982). The potential for the use of *P. punctiformis* is also being studied in Denmark, where it is native (Julien, 1982). Another accidental introduction, *P. xanthii* (Schweinitz), has spread rapidly through eastern Australia, where, in conditions favourable for disease development, it has been reported to kill creeping thistle seedlings and retard the growth and seed production of established plants (Julien *et al.*, 1979).

Leafy spurges, in particular *Euphorbia esula* L. and *E. waldstainii* (Sojak) A. Radcliffe-Smith, are serious range weeds in North America. These alien species are difficult to control with herbicides and the low value of the infested lands makes biological control an economic necessity. *Uromyces striatus* (Schroer) already infests leafy spurge but shows little potential as a biological control agent (Littlefield, 1984). Several European rusts have been tested against *Euphorbia* in the USA, and species of *Melampsora* and *Uromyces* have shown some potential value in control of *E. esula-virgata* L. and *E. cyparassias* L. (Turner & Bruckart, 1983; Sutker & Bruckhart, 1984). *U. alpestris* Transoch. and *U. scutellatus* (Pers.) Lev. s. str. are systemic rusts which show good potential as agents of control of *E. cyparissias*, the cypress spurge. Shoots are deformed and reduced in number (Defago *et al.*, 1985).

Other economically important weeds which are subjects of biological control using fungi include water hyacinth, yellow nutsedge and giant ragweed. Considerable research effort has been expended on control of water hyacinth, *Eichnornia crassipes* (Mart.) Solms, mostly using mycoherbicides. One rust fungus, *Uredo eichnornia* Gonz.-Frag. & Cif., which has been found in Argentina, is being considered for release in Florida, USA, as a classical biocontrol agent (Charudattan *et al.* (1981). Control of giant ragweed, *Ambrosia trifida* L., using *Puccinia xanthii* Halst has been proposed using isolates from North America (Barta, 1981). Giant ragweed has been introduced into Europe and the rust fungus could be of help there. The control of yellow nutsedge, *Cyperus esculentis* L. (one of the world's worst weeds), using *P. canaliculata* (Schw.) has also been proposed (Calloway *et al.*, 1985). This weed is extremely costly to control and a biocontrol agent would be welcomed. Unfortunately, the usefulness of *P. canaliculata* is questionable because it also infects the sunflower. Further work will be required to determine the degree of susceptibility of the large-seeded commercial varieties of sunflower.

6.4 Conclusion

Although classical biological control is still primarily the province of the entomologist, significant progress has been made in the classical control of weeds with pathogenic fungi in the past 20 years. The examples cited here confirm the viability of the concept. Progress is slow compared with other weed control tactics, but this is offset to a degree by the relatively minor costs and research commitment necessary in comparison with that incurred in the development and usage of chemical herbicides.

Plant pathologists are, of course, largely concerned with the *protection* of crops from disease, and only a few have pioneered the avenues of research outlined in this chapter. There are also more incentives for the mycoherbicide approach to weed control with fungi because these products are patentable and marketable whereas classical fungal agents are probably not. Until more research support is made available from public sources, research on the classical biological control of weeds with fungi may continue to lag behind other methods.

References

Adams, E.B. & Line, R.F. (1984a). Biology of *Puccinia chondrillina* in Washington. *Phytopathology*, **74**, 742–5.

Adams, E.B. & Line, R.F. (1984b). Epidemiology and host morphology in the parasitism of rush skeletonweed by *Puccinia chondrillina*. *Phytopathology*, **74**, 745–8.

Barta, W.T. (1981). *Puccinia xanthii* forma specialis *ambrosia-trifidae*, a macro-cyclic rust for the biological control of giant ragweed-, *Ambrosia trifida* (Compositae). *Phytopathologia*, **74**, 742–5.

Bruckart, W.L. Politis, D.J. & Sutker, E.M. (1985). Susceptibility of *Cynara scolymus* L. (artichoke) to *Puccinia carduorum* Jacq. observed under greenhouse conditions (Abstr.) *Proceedings of the VIth International Symposium on the Biological Control of Weeds*, Vancouver, Canada, 19–25 August 1984, ed. E.S. Delfosse, pp. 603–7. Agriculture Canada, Vancouver.

Bruzzese, E. (1984). Occurrence and spread of *Phragmidium violaceum* on blackberry (*Rubus fruticosus*) in Victoria, Australia (Abstr.). *VIth International Symposium, Biological Control of Weeds*, Vancouver, Canada.

Calloway, M.B, Phatak, S.C. Wells, H.D. (1985). Studies on alternate hosts of the rust *Puccinia canaliculata*, a potential biological control agent for nutsedges. *Plant Disease*, **69**, 924–6.

Charudattan, R., McKinney, D.E. & Hepting, K.T. (1981). Production, storage, germination and infectivity of uredospores of *Uredo eichnorniae* and *Uredo pontederia*. *Phytopathology*, **71**, 1203–7.

Cullen, J.M. (1976). Evaluating the success of the programme for biological control of *Chondrilla juncea* L. In: *Proceedings of the IVth International Symposium on Biological Control of Weeds*, ed. T.E. Freeman, pp. 117–21. University of Florida, Gainesville, FL.

Cullen, J.M. (1985). Bringing the cost benefit analysis of biological control of *Chondrilla juncea* up to date (Abstr.). *Proceedings of the VIth International Symposium on the Biological Control of Weeds*, Vancouver, Canada, 19–25 August 1984, ed. E.S. Delfosse, pp. 145–52. Agriculture Canada, Vancouver.

Cullen, J.M., Kable, P.F. and Catt, M. (1973). Epidemic spread of a rust imported for biological control. *Nature, London*, **244**, 462–4.

Defago, G., Kern, H. & Sedlar, L. (1985). Potential control of weedy spurges by the rust *Uromyces scutellatus*. *Weed Science*, **33**, 857–64.

Emge, R.G., Melching, J.S. & Kingsolver, C.H. (1981). Epidemiology of *Puccinia chondrillina*, a rust pathogen for the biological control of rush skeletonweed in the United States. *Phytopathology*, **71**, 835–43.

Field, R.P. (1985). Biological control of blackberry — resolving a conflict in Australia (Abstr.) *Proceedings of the VIth International Symposium on Biological Control of Weeds*, Vancouver, Canada, 19–25 August 1984, ed. E.S. Delfosse, pp. 341–9. Agriculture Canada, Vancouver.

Freeman, T.E., Charudattan, R. & Conway, K.E. (1978). Status of the use of plant pathogens in the biological control of weeds. *Proceedings of the IVth International Symposium on the Biological Control of Weeds*, ed. T.E. Freeman, pp. 201–6. University of Florida, Gainesville, FL.

Goeden, R.D. & Kok, L.T. (1986). Comments on a proposed 'new' approach for selecting agents for the biological control of weeds. *Canadian Entomologist*, **118**, 51–8.

Halstead, B.D., (1894). Weeds and their most common fungi. *M.J. Experimental Station Report*, 379.

Hasan, S. (1970). The possible control of skeletonweed, *Chondrilla juncea* L. using *Puccinia chondrillina* Bubak & Syd. *Proceedings of the IVth International Symposium on the Biological Control of Weeds*, ed. T.E. Freeman, pp. 11–14. University of Florida, Gainesville, FL.

Hasan, S. & Wapshere, A.J. (1973). The biology of *Puccinia chondrillina*, a potential biological control agent of skeleton weed. *Annals of Applied Biology*, **74**, 325–32.

Hokkanen, H. & Pimmental, D. (1984). New approach for selecting biological control agents. *Canadian Entomologist*, **116**, 1109–21.

Julien, M.H. (1982). *Biological Control of Weeds: a World Catalogue of Agents and their Target Weeds*. Commonwealth Agricultural Bureaux, Slough, UK.

Julien, M.H., Broadbent, J.E. & Matthews, N.C. (1979). Effects of *Puccinia xanthii* on *Xanthium strumarium* (Compositae). *Entomophaga*, **24**, 29–34.

Lee, G.A. (1986). Integrated control of rush skeletonweed (*Chondrilla juncea*) in the western US. *Weed Science*, **34**, (Suppl. 1), 2–6.

Littlefield, L.J. (1985). Biocontrol of leafy spurge with pathogenic fungi (Abstr.). *Proceedings of the VIth International Symposium on the Biological Control of Weeds*, Vancouver, Canada, 19–25 August 1984, ed. E.S. Delfosse, p. 633. Agriculture Canada, Vancouver.

Mortensen, K. (1985). Reaction of safflower cultivars to *Puccinia jaceae*, a potential biocontrol agent for diffuse knapweed (Abstr.). *Proceedings of the VIth Internationl Symposium on the Biological Control of Weeds*, Vancouver, Canada, 19–25 August 1984, ed. E.S. Delfosse, pp. 447–52. Agriculture Canada, Vancouver.

Nakao, H.K. & Funasaki, G.Y. (1979). Introductions for biological control in Hawaii: 1975 and 1976. *Proceedings of the Hawaiian Entomological Society*, **13**, 125–8.

Oehrens, E. (1977). Biological control of the blackberry through the introduction of rust, *Phragmidium violaceum*, in Chile. *Food and Agricultural Organisation Plant Protection Bulletin*, **25**, 26–8.

Oehrens, E.B. & Gonzales, S.M. (1974). Introduction de *Phragmidium violaceum* (Schultz) Winter como factor de control biologico de zarzamora (*Rubus constrictus* Lef. et M. y *R. ulmifolius* Schott.). *Agro Sur*, **3**, 87–91.

Oehrens, E. & Gonzales, S. (1975). Introduction de *Uromyces galegae* (Opiz) Saccardo como factor de control biologico de galega (*Galega officinalis* L.). *Agro Sur*, **3**, 87–91.

Oehrens, E. & Gonzales, S. (1977). Dispersion, ciclo biologico y danos causados por *Phragmidium violaceum* (Schultz) Winter en zarzamora (*Rubus constrictus* Lef. et M. y *R. ulmifolius* Schott.) en las zonas centro-sur y sur de Chile. *Agro Sur*, **5**, 73–85.

Olivieri, I. (1984). Effect of *Puccinia cardui-pycnocephali* on slender thistles (*Carduus pycnocephalus* and *C. tenuifloris*). *Weed Science*, **32**, 508–10.

Politis, D.J., Watson, A.K. & Bruckart, W.L. (1984). Susceptibility of musk thistle and related composites to *Puccinia carduorum*. *Phytopathology*, **74**, 687–91.

Schmidl, L. 1980. Some aspects of problems arising with the biological control of skeletonweed, *Chondrilla juncea* L., in Victoria. *Third Australian Applied Entomological Research Conference*, Lowes, Queensland, Working Papers 2(b), 1–2.

Sutker, E.M. & Bruckart, W.L. (1984). Host specificity of *Melampsora euphorbiae*, a pathogen of *Euphorbia cyparissias* (Abstr.). *Phytopathology*, **74**, 864.

Templeton, G.E. (1982). Status of weed control with plant pathogens. In: *Biological Control of Weeds with Plant Pathogens*, eds R. Charudattan & H.L. Walker, pp. 29–44, Wiley, New York.

Templeton, G.E. & Trujillo, E.E. (1981). The use of plant pathogens in the biological control of weeds. In: *CRC Handbook of Pest Management in Agriculture*, ed. D. Pimentel, pp. 345–50, CRC Press, Boca Raton, FL.

Trujillo, E.E. & Obrero, F.P. (1972). The biological control of *Cassia surattensis*, a brush weed of pastures in Hawaii with *Cephalosporium* sp. (Abstr.). *Phytopathology*, **62**, 793.

Trujillo, E.E. & Obrero, F.P. (1978). *Cephalosporium* wilt of *Cassia surattensis* in Hawaii. *Proceedings of the IVth Symposium on the Biological Control of Weeds*, ed. T.E. Freeman, pp. 217–20, University of Florida, Gainesville, FL.

Turner, S.K. & Bruckart, W.L. (1983). European rust fungi pathogenic to collections of leafy spurge from the United States. (Abstr.). *Phytopathology*, **73**, 969.

Turner, S.K., Kwaitkowski, A. & Fay, P.K. (1982). Stimulation of *Puccinia obtegens* teliospore germination (Abstr.). *Phytopathology*, **72**, 711.

Wapshere, A.J. (1979). Recent progress in the biological control of weeds. *European and Mediterranean Plant Protection Organisation*, Bulletin 9 (1), 95–105.

Watson, A.K. (1985). Host specificity of plant pathogens in biological weed control (Abstr.). *Proceedings of the VIth International Symposium on the Biological Control of Weeds*, Vancouver, Canada, 19–25 August 1984, ed. E.S. Delfosse, pp. 577–586. Agriculture Canada, Vancouver.

Watson, A.K. & Brunetti, K. (1984). *Puccinia carduorum* on *Carduus tenuiflorus* in California. *Plant Disease*, **68**, 1003–5.

Watson, A.K., Schroeder, D. & Alkhoury, I. (1981). Collection of *Puccinia* species from diffuse knapweed in Eastern Europe. *Canadian Journal of Plant Pathology*, **3**, 6–8.

Wilson, C.L. (1969). The use of plant pathogens in weed control. *Annual Review of Plant Pathology*, **7**, 411–33.

Competitive antagonism of soil-borne plant pathogens

7.1 Introduction

Most of the early studies on host–parasite relationships remained firmly devoted to examining the parasite in a microbiological vacuum, and no heed was paid to the other microbial occupants of an infection court. It was not until 1926 when Sanford suggested that the alleviation of symptoms of potato scab (*Streptomyces scabies*), achieved by the incorporation of grass cuttings into soil, was due to a biological control that interest in the microbial ecology of plant pathogens increased. There followed a rush to demonstrate biological control of many different diseases, often in plate and pot tests where sterile, semi-sterile or heavily organically amended soil was used as a planting medium. The promising results from these laboratory tests were never fulfilled in the subsequent field trials, because these initial field screenings disregarded the stable structure of the soil microbial community in which the microbial buffering power prevented the establishment of newcomers that were not ecologically suited to the new environment. We now know that any attempt to boost the numbers of a population of an antagonistic microorganism by its introduction alone is doomed to failure, because, as Odum (1971) stated, populations are a reflection of habitat, and any change due to introduction without change in habitat must be a transient one. Many examples of such experiments are cited in Baker & Cook (1974), Papavizas & Lewis (1980) and Cook & Baker (1983). Thus inappropriate laboratory and small-scale field tests have led to increasing numbers of examples of failed mass introduction of microbes into unmodified field soils. This in turn has led to a considerable misjudgement of the potential of biological control of soil-borne plant pathogens (Schroth & Hancock, 1981). The knowledge that environmental changes are necessary to allow introduced antagonists to become established seems to have had very little effect on the methodology researchers have used in field experiments The ecological approach recommended by Garrett in 1956 has only recently been accepted as providing the best way forward, and we are now seeing more successful examples of biological control of root disease in field conditions.

7.2 Competitive antagonism

In natural systems, competition, antibiosis and hyperparasitism play an important role in determining the outcome of a relationship between two microorganisms (Lumsden, 1981; Singh & Faull, 1986). The features of the first two categories will be discussed in turn. Hyperparasitism is dealt with by Blakeman in Chapter 8, this volume. In order to understand why there have been so many failures in the biological control of soil-borne plant pathogens by mass introduction of antagonists, and to offer evidence of better approaches to this problem, it is necessary also to consider in detail the ecology of the soil and rhizosphere environment.

7.2.1 *Competition*

Just as competition for light is the most important factor in the evolution and sociology of higher plants, so competition for substrates is the most important factor for heterotrophic soil fungi (Garrett, 1956). Soil fungi require water, oxygen and nutrients derived from the living or dead soil flora and fauna. Clark (1965) claimed that competition for water between microorganisms does not occur because its availability is determined by water potential, which microorganisms are unable to affect significantly. It has also been asserted that fungi would tend to produce water by their activities rather than deplete it (Frankland, 1981). It does seem likely, however, that in microenvironments in soil, like those that occur in the water films in soil pores, microorganisms would be able to alter the water potential by its utilisation. Tolerance of low water potentials and specialisations that overcome low water potential can enable some microorganisms to exploit marginal environments. For example, *Aspergillus glaucus* spores have the ability to germinate in grain containers where the relative humidity is 73% (Griffen, 1963) and on wheat grains with a moisture content as low as 14.5% (Christensen, 1957).

Competition for oxygen is likely to be important too, in that diffusion of oxygen through water in soil is slow, and for some periods of the year soils can remain very wet. Greenwood and Goodman (1964) found that anaerobiosis was common in microsites in soil and that these microsites were most often distributed in the top few centimetres of soil. Thus tolerance of microaerophyllic conditions may confer ecological advantages on some microbes.

Mineral nutrients are normally present in such large quantities in soil environments that competition for them is unlikely to occur. The exception to this may be nitrogen. Competition for nitrogen can occur in substrates were nitrogen content is low, and some fungi seem physiologically specialised for growth in such habitats, where they may have an ecological advantage (Lockwood, 1981).

Carbohydrate substrates are discontinuous in space and time. For example, the rhizosphere, which represents one of the most active microbiological zones below soil, is spatially very transient. Its highest activity is around the young actively growing root tip, and as the root tip grows onwards through the soil so the most active rhizosphere zone moves with it.

Within each niche in the soil, substrates are available in different quantities for varying lengths of time with the result that communities of microbes are in a state of constant change and succession. The soluble substrates such as sugars and amino acids are utilised by the sugar fungi, whereas more intractable substrates such as cellulose and lignin are decayed by other slower growing fungi that have to produce complex combinations of extracellular enzymes to decompose such substrates at considerable energy costs. This succession is much like that which occurs in higher plant communities, the dominant preparing the way for their own replacement through the changes they produce in the substrate and the habitat. However, unlike higher plant communities, this succession leads finally to the depletion of the substrate, and a theoretical end point of zero (Garrett, 1956; Frankland, 1981).

The characteristics of the different nutritional groups that exist in the soil vary considerably. The fast-growing Phycomycetes are virtually completely dependent upon soluble sugars as carbohydrate sources. None can utilise lignin and very few can utilise cellulose, but their very rapid assimilation of the soluble carbohydrate resources allows them to germinate, grow and sporulate in the relative absence of competition from other nutritional groups of soil fungi.

Substrates like cellulose and lignin are degraded by groups of fungi that belong to the Ascomycetes, Basidiomycetes and the Deuteromycetes. It is difficult to assess the importance of the members of the Basidiomycetes in terms of substrate utilisation in field soils as they are difficult to estimate in number and bulk, but certainly the Ascomycetes and the Deuteromycetes have an important role to play in this environment. The preferential establishment of different species of these fungi on cellulosic and lignin-containing substrates is dependent on several factors. The inoculum potential of each species is important; those fungi with the highest numbers of propagules or the greatest mass of mycelial growth will have the greatest competitive advantage. Thus the occupation of a previous substrate that has allowed the accumlation of great numbers of spores or mycelium confers advantages in the colonisation of new substrates. A second factor is that the complex substrate colonisation will be influenced by the continuing microbial activity of the early colonists, especially by the Phycomycetes. The production of CO_2, utilisation of O_2, acidification of the substrate, and production of NH_3 and antibiotics by these fungi all have an effect on microbial succession (Connell & Slayter, 1977; Lockwood, 1981).

7.2.2 *Antibiosis*

Much controversy has centred around the significance of the production of antibiotics by microorganisms in terms of survival value and their role in competition and substrate possession. Arguments against their importance have suggested that they may not be produced in large enough quantities in soil to be significant because sufficient nutrients for their production are not present (Gottlieb, 1976). Even when they are produced they may be bound to soil particles, particularly clay micelles, and inactivated (Pinck *et al.*, 1961a, b; Soulides *et al.*, 1962) or they may be decomposed by other microorganisms (Jeffries, 1952; Brian, 1957) Another argument against the significance of antibiotic production in soil is that microorganisms that possess the ability to produce them are not spectacularly more successful that those that do not (Gottlieb, 1976). Counter-arguments to these are based on the suggestion that it is quite possible that in microenvironments where fresh substrates are being colonised concentrations of antibiotics are likely to be high enough to exert a selective effect on other microorganisms and ensure substrate possession by their producers (Rothock & Gottlieb, 1984). Also, Park (1967) has suggested that the determination of gross quantities of antibiotics in soil is likely to have very little relation to the concentration in local environments. The fact that antibiotic producers are not exclusively successful merely reflects the fact that antibiosis is only one of several mechanisms of competition between microorganisms, and in other circumstances other mechanisms are likely to be more important. The greatest amount of evidence for antibiosis as a significant phenomenon in microbial interactions in soil is indirect, and has been summarised by Brian (1957). Brian concludes that it is only on carbohydrate-rich substrates that competition is extreme, and organisms isolated from such substrates are often able to produce antibiotics. The information gathered by Brock (1966) supports the hypothesis that antibiotics are active in nature, and that they may be of great importance in the process by which a secondary colonist might invade a substrate that had already been partially depleted of nutrients and already occupied by a primary colonist (Wicklow, 1981).

The fungi have been demonstrated to produce a very wide variety of toxic substances that have activity against a range of eukaryotic organisms. In the colonisation of some substrates, such as seeds, fungi have been shown to have the ability to produce such large quantities of extracellular products that they constitute a toxic hazard to men and animals that eat them (Smith & Moss, 1985) However, large quantities of these antibiotics are not produced during normal vegetative growth because of internal regulation. The production of these secondary metabolites occurs in circumstances where the mycelial growth of the fungus has ceased owing to some nutrient being limiting (Demain, 1972; Moss, 1984), often while there is still an ample carbon source. In these circumstances, metabolism switches to the secondary pathways that produce antibiotics (Bu'lock *et al.*, 1974; Moss, 1984).

Antibiosis can be envisaged as having advantages over other types of antagonism for biological control. The antibiotic substance may diffuse in water, air or substrate to other microorganisms, and thus direct contact between the two is not necessary. The exact distance any antibiotic can travel is determined by its chemical characteristics and by the characteristics of the carrier matrix. The effect it will have on any target microorganisms will depend on the concentration of the antibiotic at that point, and on the susceptibility of the target organism to it. Nevertheless, the sphere of influence of an antibiotic-producing colony is likely to extend several times over its physical volume. However, it may be counter-productive to release excessive amounts of antibiotic, because the lysis of targets may then occur beyond a zone in which the antagonists can retrieve the newly available nutrients with the result that the released nutrients could be used by potential competitors (Wicklow, 1981).

The ability of a fungus to produce an antibiotic may thus be very important in determining its ability to colonise and to maintain its presence on a substrate. Bruehl *et al.* (1969) showed that *Cephalosporium graminearum* produced a wide-spectrum antibiotic that enabled it to retain possession of a substrate for 2 to 3 years, whereas non-antibiotic-producing strains of the same organism were overrun on the same substrate by saprophytes within a few months. However, the extent to which the antibiotic-producing fungus may colonise beyond its substrate may be limited. Sewell (1959), for example, showed that a *Penicillium* sp. colonised small individual substrates in soil but did not extend beyond those substrates into the surrounding soil. Nevertheless, antibiotics may still be effective even in these limited niches. Wright (1956a, b) demonstrated that colonisation of pea seeds by *Trichoderma viride* (possibly *Gliocladium virens*) resulted in the accumulation of significant amounts of the antibiotic Gliotoxin in the seeds, and that when *Trichoderma* was used as seed-coat inoculant it was able to control *Pythium ultimum* Thom (Wright, 1956c).

Equally important is the relative sensitivity of fungal species to toxins secreted by other species of soil microorganisms. It appears that those fungi that produce antibiotics are more likely to be resistant to them (Jeffries, *et al.*, 1953). Antibiotic production by the initial colonisers of a substrate will thus affect subsequent colonists. Horn (1977) described in higher plant communities a competitive hierarchy, which was a type of succession in which later successional species are able to dominate earlier successional species and thus regulate the sequence of colonists in a successional sere. Such a competitive hierarchy between microorganisms would enable later species of colonist to transfer nutrient sources by lysis from the mycelium of earlier occurring, faster growing colonists (Wicklow, 1981).

7.3 Competitive antagonism and biological control

Since competition and antibiosis are crucial in establishing a fungus on a substrate, they will of course determine which species will occupy a substrate and which will not. It is the possession of a substrate that is vital to the establishment of a biological control agent in any environment. Once established within a niche, the disease antagonist can prevent the ingress of the pathogen for as long as that niche exists. However, as resources are depleted, so the total niche space is reduced and competition for any remaining resource will increase (Swift, 1976). It is known that the persistence of fungi in natural soil depends on a regular input of new substrates. Thus introduced antagonists will only have their initial substrate for a brief period of time, and it is therefore extremely important that the composition of the substrate on which any microbial inoculum is grown and applied be carefully formulated. The incorporation of inoculum of biological control agents into substrates containing large quantities of soluble sugars may merely provide a substantial stimulus to the Phycomycetes. Most antagonists that have been introduced are not members of this nutritional group, and grow more slowly. After the Phycomycetes have rapidly utilised the soluble substrates, the inoculant has to compete for other substrates with indigenous groups of cellulose and lignin decomposers that already have a niche in that environment and probably have greater inoculum potential than the introduced species. Even if the inoculant does germinate, the intense competition for nutrients would render the germling susceptible to the germination/lysis phenomenon (Ko & Lockwood, 1970).

The formulation of disease antagonists on more complex substrates may also prove to be unsatisfactory. The decomposition of cellulose and lignin involves a considerable expenditure of energy for the production of extracellular enzymes such as cellulases and ligninases. Thus the provision of soluble sugars may be necessary to permit germination and initial growth before extracellular enzyme production can begin (Norkrans, 1950; Lockwood, 1977).

The pattern of distribution of an antagonist on a substrate is also important. If the colonisation is superficial, much of the substrate will remain uncolonised and potentially available to other microbes, whereas a more thorough colonisation of a substrate will prevent this occurring. This is particularly important when a biological control agent is prepared using solid substrate fermentation. Incomplete utilisation of the substrate may be caused by other environmental factors within the substrate which are sub-optimal for growth of the organism concerned, such as gas exchange or water.

Even if the introduced antagonist is able to utilise the soluble carbohydrate before the resident soil sugar fungi, there remains the problem of subsequent competition for more complex substrates in the inoculum, such as cellulose. The outcome of the interactions will depend heavily on the relative inoculum potentials of the two protagonists, and on their competitive saprophytic ability. Garrett (1956) defined competitive saprophytic ability as the summation of physiological characteristics that make for success in the colonisation of dead organic substrates. High competitive saprophytic ability depends on five characteristics: rapid spore germination, high growth rate, high extracellular enzyme production, and the production and tolerance of antibiotics. The environment around the substrate will affect all of these characteristics.

It is relatively simple to ensure that the number of antagonist propagules in a soil is higher than the number of propagules of any other fungus, but ensuring that all other factors are favourable is more difficult. Growth rates and rates of spore germination are characteristics that are genetically determined in the antagonist, and it is often difficult to alter such characteristics. They are also considerably affected by the substrates on which they grow. Similarly the production of extracellular enzymes and toxins and tolerance to toxins produced by other microorganisms are also factors that are genetically determined.

The factors that influence the above characters will now be discussed, and the possibilities for conventional and genetic engineering techniques aimed at improving such characters will also be considered.

7.3.1 *Rapid spore germination and high growth rate*

Microbes have only a very limited capacity for controlling their micro-environment. In order to survive and compete successfully with other microbes under varying conditions, they have evolved the potential to respond to environmental change by changing themselves structurally and functionally. Adaptation can occur either by changes in the genetic constitution or by phenotypic adjustment of a given constitution. Organisms may be highly versatile, or they may be very restricted in their tolerance of differing environmental conditions. The more versatile organisms do not express their entire genome under one set of environmental conditions (Koch, 1976; Harder *et al.*, 1984) but use only that part that is structurally and functionally adjusted to the prevailing conditions. Those mechanisms that adapt structurally and functionally to an environment ensure the prolonged survival of the organism and enable it to grow as fast as possible. The mycelial form of a fungus is an ephemeral one, vulnerable to many abiotic (desiccation, UV light, mechanical damage) and biotic (microbial competition, grazing, substrate exhaustion) factors by nature of its hyaline, thin-walled, rigid hyphae. The most frequent method by which fungi survive adverse conditions is by the production of one or more types of thick-walled bodies such as chlamydospores and zygospores, or the production of aggregates of hyphae such as sclerotia and rhizomorphs. The resting spores may remain deeply dormant for long periods and have their reserves in the form of oil droplets. They require a period of after-ripening before they can germinate. The ecological usefulness of these resting spores is to some extent offset by their relatively poor dispersal as they are usually large and dense and are often formed embedded in other vegetative material. In additon, the depth of dormancy that some of these resting spores enter makes them unsuitable as inoculum for biological control agents. The zoospores and sporangiospores of the lower fungi, the conidia of the Ascomycetes and the urediospores of the rust fungi are all designed primarily as dispersal propagules rather than resting spores and they serve to disseminate the fungus over a suitable substrate when other environmental conditions are favourable. As such, neither spore type is ideally suited as inoculum, and a compromise of a combination of spore types is often needed.

The growing fungus also has various responses to these adverse environmental conditions. The lower fungi, such as species of *Mucor*, have the ability to grow very rapidly and thus escape adverse conditions such as substrate depletion. Other soil fungi (e.g. *Penicillium*, *Aspergillus* and *Trichoderma*) that are numerically as successful as *Mucor* have slow growth rates, but this is compensated for by the formation of very large numbers of spores, which are easily dispersed to the surrounding areas to be followed by equally rapid germination (Hawker, 1957). Most spores are rapidly attacked and lysed in soil unless they are protected by large amounts of organic matter (Park, 1955) and thus rapid germination is an essential part of survival of adverse conditions.

In summary, therefore, rapid growth, profuse sporulation, efficient dispersal of propagules and the ability to rapidly colonise new substrates will enable a fungus to exploit favourable habitats and to survive by migration when environmental conditions are no longer favourable. Such a combination of characteristics would make for an ideal biological control agent.

However, many fungi do not have all or indeed any of these characteristics, but they survive through having different strategies, such as the exploitation of substrates that others cannot exploit, or by the production of extracellular products such as antibiotics and secondary metabolites that make an environment unfavourable to other organisms. Often the habitats in which these chemicals are produced are not originally borderline ones, but when colonised by an antibiotic producer the production of these antibiotics renders the habitat borderline to all others (Brian, 1957; Hawker, 1957). Tolerance to pesticide and other chemical residues can also allow certain fungi to recolonise semi-sterile areas. Species of *Trichoderma*, for example, are resistant to low doses of soil fumigants and can recolonise soil and wood after such treatment (Warcup, 1951; Evans, 1955).

Most fungi have very similar growth optima, and it is the tolerance of conditions that are sub-optimal that allows certain fungi to colonise otherwise empty habitats. For example, acid tolerance, sugar tolerance and temperature tolerance allow fungi to colonise oranges, jam and compost heaps, respectively. Plant-pathogenic fungi tolerate the sub-optimal conditions within plant tissues which effectively exclude competitors. Thus, ideally, the ability to exploit unusual substrates or tolerate extreme environments would give a biological control agent an ecological advantage over other competitive soil microorganisms.

7.3.2 *Extracellular-enzyme and antibiotic production*
Primary metabolism is basically the same for all living systems, and involves an interrelated series of enzyme-mediated catabolic, amphibolic and anabolic reactions, which provide biosynthetic intermediates and energy and convert biosynthetic precursors into macromolecules such as RNA, DNA, proteins and polysaccharides. Primary metabolism is finely balanced, and intermediates rarely accumulate. In contrast, secondary metabolites are produced subsequent to growth and they have no function in growth though they may be important in survival. Often they are restricted to certain taxonomic groups, have unusual chemical structures and are often present and synthesised in complex family groups. It is during the active

phase of primary metabolism that extracellular enzyme production takes place, whereas it is during secondary metabolism that large quantities of antibiotics and organic acids are synthesised.

During normal metabolism an organism will produce small quantities of hydrolases, particularly under growth limitation by carbon and nitrogen sources. These hydrolases are released into the environment where they act as 'scouting molecules'. If a substrate for such an enzyme is available in the microenvironment of the organism, it will slowly be hydrolysed and the products of hydrolysis may diffuse back to the cell, thus signalling the presence of a usable substrate. These products then in turn act as inducers for the synthesis of more enzymes. Just as low levels of catabolites switch on enzyme synthesis, so excess amounts of catabolites switch off enzyme production, and so extracellular enzyme production is regulated by catabolite repression (Koch, 1976). Thus natural or artificially constructed strains of microorganisms that are less sensitive to feedback inhibition should give enhanced yields of these extracellular products. Greatest research has been concerned with the overproduction of extracellular enzymes that cause hydrolytic reactions, and it is from this work that general inferences have to be made. It is fair to say, however, that techniques for achieving maximum extracellular enzyme production by a microorganism are rather empirical at the very best, and more will be said on such techniques in the next section.

The selection of isolates for extracellular enzyme production is based on the selection and mutagenesis of originally wild strains that are known to produce the metabolite required. Strain selection will involve the isolation and testing of many strains of the organism, capitalising on natural genetic variation (Johnston, 1977). The principles and techniques of isolation of microorganisms from their habitats have been summarised by Starr *et al.* (1981) for bacteria, by Williams and Wellington (1982) for Actinomycetes and by Booth (1971) for the fungi. Mutation and selection can utilise techniques such as the use of ionising radiation, nitrous acid, base analogues or alkylating agents. Some methods cause chromosomal damage and can lead to problems in the stability of the mutant. The methodology for these techniques has been outlined by Alder (1971) and Calan (1970). The selection of improved yield mutants is more difficult than the selection of auxotrophic, colour or resistance mutants, and screens are frequently conducted involving only one unreplicated flask. Unless enhanced yield relative to control is observed from that flask, then the mutant is discarded. Recombination and selection techniques may involve the hybridisation of isolates created by mutation with old or new strains. For example, *Penicillium chrysogenum* WC-9, a vigorous diploid strain, was created as a genomic duplication of the original haploid strain (Pathak and Elander, 1971). Most strain selection programmes have been run on such a system of successive rounds of intense mutagenesis followed by screening, with no knowledge of the biochemical changes required to produce such improved strains. It is interesting to note that at present the contribution of genetic engineering to these programmes is zero (Hopwood & Johnston, 1984) although this is quite likely to change in the near future.

There are often serious setbacks to these programmes; in particular improved strains can lose their desired properties by a process known as strain degeneration. This is because the environment in which the improved strain grows must not select against it or the mutation will become a genetic burden. The improved strain behaviour must offer a selective growth advantage or strain degeneration will become strain improvement in terms of survival for the organism (Neijssel & Tempest, 1979). It has to be remembered that an organism will strive to remain alive and to create progeny (Darwin, 1859), and survival rather than maximal yield with respect to every nutrient is the first priority. A good example of this can be found in the behaviour of *Escherichia coli* in culture. In aerobic fermentation *E. coli* cells multiply rapidly, producing carbon dioxide, but during anaerobic fermentation low numbers of cells are produced, together with quantities of ethanol, succinate, lactate, acetate, H^+ and CO_2 because of the metabolic restrictions that the absence of oxygen causes. In order to allow for metabolism and growth under these conditions, the cell must dispose of reducing equivalents. The accumulated intermediates of metabolism function as acceptors for these and they are excreted, together with potentially toxic metabolites (Neijssel & Tempest, 1979).

It can thus be seen that, if environments could be constructed in which it was essential for an organism to excrete metabolites in order to maintain viability and grow, metabolite production would be stable and the reproducible property of the isolate and strain degeneration would not occur. Culture conditions that will promote metabolite production have been reviewed by Pirt (1969), Christner (1976) and Tempest & Neijssel (1978). However, these exacting conditions are requirements that are known to exist for the sterile bulk fermentation of microbes with maximum extracellular metabolite production, and as such are attainable. It is not possible to achieve such a level of control over the physiology of microbes in the rhizosphere, phyllosphere or bulk soil, and thus the environmental manipulation that can prevent strain degeneration under synthetic culture conditions cannot occur. Thus any microbial control agent that produces excess extracellular compounds to the detriment of general physiology will tend to be under selective pressure to reduce output. Any newly constructed isolate will have to be even more environmentally fit than the wild type. Therefore although it is possible to pinpoint changes that could be made in the physiology of a potential biological control agent that would improve individual facets of its competitive saprophytic ability, it is more difficult to envisage these changes improving the whole of its competitive saprophytic ability.

7.4 Conclusions

The basic flaw in using competitive antagonism to control soil-borne plant pathogens lies in the very considerable structural, physical and physiological requirements we ask them to fulfil. A perfect sub-soil biological control agent would have to be living, physiologically active, producing extracellular antibiotics and enzymes over a period of weeks if not months, through an environment that will be continuously changing in nutritional status, water potential and microbial population. It is these demanding requirements that have led to so many failures of biological control in the field. To achieve a greater degree of success we must examine each requirement in great detail in the light of each particular antagonist's physiology and biology, and accept the limitations of the organism. We must determine which spore type or mix of spore types will be most suitable as inoculum for one condition and which substrate will give rise to exclusive utilisation by the biocontrol agent. Should the antagonist produce an antibiotic, we must ascertain the conditions that will maximise its production, so that the inoculum that is placed in the field thoroughly impregnated with it. In this way other competitive organisms will be unable to possess the substrate and dislodge the antagonist. We must also understand that genetic engineering may not hold all the answers, as alteration of the antagonist's genome to increase biological control potential may have some physiological consequences that place the antagonist at a disadvantage in the environment which could lead ultimately to strain degeneration.

Advances in the use of microbes as control agents of soil-borne plant pathogens will no doubt be made, but it is worth remembering that the chemical industry with its years of expertise and multi-million pound research programmes have as yet failed to improve upon the percentage of crops that are lost due to pests, weeds and diseases (Faull, 1986). Soil-borne plant pathogens are extremely well adapted to their niche, and the successful examples of induced biological control of these diseases rely on three special environmental conditions: soil sterilisation (Munneke *et al.*, 1981), heavy soil amendment (Lumsden *et al.*, 1983) or large-scale soil transfer (Shipton *et al.*, 1973). The natural examples of biological control of these diseases are situations where disease decline has been slowly achieved over years of soil husbandry including green manuring shallow ploughing and careful selection of crop rotations or in some cases of crop monoculture. The phenomena of disease-suppressive soil is still not fully understood but again is something that takes years to develop (Cook & Baker, 1983). One of our most exacting requirements for a biological control agent is the time scale over which we expect it to work and the speed at which it must have a visible effect. Biological control of soil-borne plant pathogens may well be possible, but it will be only one part of an integrated pest management scheme that will involve crop and soil husbandry, crop rotation and the use of disease-resistant crop plant species. Given a little patience, at least some of systems could work.

References

Alder, H.I. (1971). Techniques for the development of novel micro-organisms. In: *Radiation and Radioisotopes for Industrial Micro-organisms*, pp. 241–9. International Atomic Energy Agency, Vienna.

Baker, K.F. & Cook, R.J. (1974). *Biological Control of Plant Pathogens*. W.H. Freeman, San Francisco.

Booth, C. (1971). Fungal culture media. In: *Methods in Microbiology 4*, ed. C. Booth, pp. 49–94. Academic Press, New York.

Brian, P.W. (1957). The ecological significance of antibiotic production In: *Microbial Ecology*, eds R.E.O. Williams & C.C. Spicer, pp. 168–88. Cambridge University Press, Cambridge.

Brock, T.D. (1966). *Principles of Microbial Ecology*. Prentice Hall, Englewood Cliffs, NJ.

Bruehl, G.W., Millar, R.L. & Cunfer B. (1969). Significance of antibiotic production by *Cephalosporium graminearum* to its saprophytic survival. *Canadian Journal of Plant Science*, **49**, 235–46.

Bu'lock, J.D., Detroy, R.W., Hostalek, Z. & Munim-Al-Sharkarchi, A. (1974). Regulation of secondary biosynthesis in *Gibberella fujikoroi*. *Transactions of the British Mycological Society*, **62**, 377–89.

Calan, C.T. (1970). Improvement of micro-organisms by mutation, hybridisation and selection. In: *Methods in Microbiology 3A*, eds J.R. Norrus and D.W. Ribbons, pp. 435–59. Academic Press, New York.

Christensen, C.M. (1957). Deterioration of stored grains by fungi. *Botanical Reviews*, **23**, 108–34.

Christner, A. (1976). Spezifische Produktsbildungsraten von Mikrobiellen Produktsbildern. *Zeitschrift für Allgemeine Mikrobiologie*, **16**, 157–72.

Clark, F.E. (1965). The concept of competition in microbial ecology. In: *Ecology of soil-borne Plant Pathogens*, eds K.F. Baker & W.C. Snyder, pp. 339–45. University of California Press, California.

Connell J.H. & Slayter, R.O. (1977). Mechanism of succession in natural communities and their role in community stability and organisation. *American Naturalist*, **111**, 1119–44.

Cook, R.J. & Baker, K.F. (1983). *The Nature and Practice of Biological Control*. American Phytopathological Society, St. Paul, Minn.

Darwin, C. (1859). *The Origin of the Species by Means of Natural Selection*. Penguin Books, London.

Demain, A. (1972). Cellular and environmental factors affecting the synthesis and excretion of metabolites. *Journal of Applied Chemistry and Biotechnology*, **22**, 345–62.

Evans, E. (1955). Survival and recolonisation by fungi in soil treated with formalin or carbon disulphide. *Transactions of the British Mycological Society*, **38**, 335.

Faull, J.L. (1986). Fungi and their role in crop protection. In: *Biotechnology and Crop Improvement and Protection*. British Crop Protection Council Monograph 34, ed. P.R. Day, pp. 141–50, British Crop Protection Council Publications, Thornton Heath.

Frankland, J.C. (1981). Mechanisms in fungal successions. In: *The Fungal Community*, eds D.T. Wicklow & G.C. Carroll, pp. 403–26. Marcel Dekker, New York.

Garrett, S.D., (1956). *Biology of the Root Infecting Fungi*. Cambridge University Press, Cambridge.

Gottlieb, D. (1976). The production and role of antibiotics in soil. *Journal of Antibiotics*, **29**, 987–1000.

Greenwood, D.J. & Goodman, D. (1964). Oxygen diffusion and aerobic respiration in soil spheres. *Journal of the Science of Food and Agriculture*, **15**, 579–88.

Griffen, D. M. (1963). Soil moisture and the ecology of soil fungi. *Biological Reviews*, **38**, 141–66.

Harder, W., Dijkhuizen, L. & Veldkamp, H. (1984). Environmental regulation of microbial metabolism. In: *The Microbe, Part 2: Prokaryotes and Eukaryotes*, Symposium of the Society for General Microbiology 36, eds D.P. Kelley & N.G. Carr, pp. 51–95. Cambridge University Press, Cambridge.

Hawker, L.E. (1957). Ecological factors and the survival of fungi. In: *Microbial Ecology*, eds R.E.O Williams & C.C. Spicer, pp. 238–58. Cambridge University Press, Cambridge.

Hopwood, D.A. & Johnston, W.B. (1984). Microbe creation by genetic engineering. In: *The Microbe, Part 2: Prokaryotes and Eukaryotes*, Symposium of the Society for General Microbiology 36, eds D.P. Kelley & N.G. Carr, pp. 257–81. Cambridge University Press, Cambridge.

Horn, H.S. (1977). Succession. In: *Theoretical Ecology*, ed. R.M. May, pp. 187–204. W.K. Saunders, Philadelphia, PA.

Jeffries, E.G. (1952). The stability of antibiotics in soils. *Journal of General Microbiology*, **7**, 295–312.

Jeffries, E.G., Brian, P.W., Hemming, H.G. & Lowe, D. (1953). Antibiotic production by the microfungi in acid heath soils. *Journal of General Microbiology*, **9**, 314–41.

Johnston, J.R. (1977). Strain improvement and strain stability in filamentous fungi. In: *The Filamentous Fungi, Vol. 1, Industrial Mycology*, eds J.E. Smith, D.R. Berry & B. Kristiansen, pp. 59–78. Edward Arnold, London.

Ko, W.H. & Lockwood, J.C. (1970). Mechanism of lysis of fungal mycelia in soil. *Phytopathology*, **60**, 148–54.

Koch, A.L. (1976). How bacteria face depression, repression and derepression. *Perspectives in Biology and Medicine*, **20**, 44–63.

Lockwood, J.L. (1977). Fungistasis in soils. *Biological Reviews*, **52**, 1–43.

Lockwood, J.L. (1981). Exploitative competition. In: *The Fungal Community*, eds D.T. Wicklow & G.C Carroll. PP. 319–59. Marcel Dekker, New York.

Lumsden, R.D. (1981). Ecology of mycoparasitism. In: *The Fungal Community*, eds D.T. Wicklow & G.C. Carroll, pp. 295–318. Marcel Dekker, New York.

Lumsden, R.D., Lewis, J.A. & Papavizas, G.C. (1983). Effect of organic amendments on soil borne plant disease and pathogen antagonists. In: *Environmentally Sound Agriculture*, ed. W. Lockerty, pp. 51–70. Praeger, New York.

Moss, M. (1984). The mycelial habit and secondary metabolite production. In: *The Ecology and Physiology of the Fungal Mycelium*, British Mycological Society

Symposia 8, eds D.H. Jennings & A.D. Rayner, pp. 127–42. Cambridge University Press, Cambridge.

Munneke, D.E., Kolbezen, M.J., Wilbur, W.D. & Ohr, H.D. (1981). Interactions involved in controlling *Armillaria mellea*. *Plant Disease*, **65**, 384–9.

Neijssel, O.M. & Tempest, D.W. 1979. The physiology of metabolite overproduction. In: *Microbial Technology, Current State, Future Prospects*. Symposium of the Society for General Microbiology 29, eds A.T. Bull, D.C. Ellwood & C. Ratledge, pp. 53–82. Cambridge University Press, Cambridge.

Norkrans, B. (1950). Studies in growth and cellulolytic enzymes of *Tricholoma*. Symposia Botanica Upsalinensis, **2**, 1.

Odum. E.P. (1971). *Fundamentals of Ecology*, 3rd edn. W.H. Saunders, Philadelphia, PA.

Papavizas, G.C. & Lewis, J.A. (1980). Introduction and augmentation of microbial antagonists for the control of soil borne plant pathogens. In: *Biological Control in Crop Protection*, ed. G.C. Papavizas, pp. 305–23. Allanheld & Osmun, Tobowa, N.J.

Park, D. (1955). Experimental studies on the ecology of fungi in soil. *Transactions of the British Mycological Society*, **38**, 130–42.

Park, D. (1967). The importance of antibiotics and inhibitory substances. In: *Soil Biology*, eds A.F. Bruges & F. Raw, pp. 435–47. Academic Press, New York.

Pathak, S.G. & Elander, R.P. (1971). Biochemical properties of haploid and diploid strains cf *Penicillium chrysogenum*. *Applied Microbiology*, **27**, 266–371.

Pinck, L.A., Holton, F.W. & Allison, F.E. (1961). Antibiotics in soil. 1. Physicochemical studies of antibiotic–clay complexes. *Soil Science*, **91**, 22–8.

Pinck, L.A., Soulides, D.A. & Allison, F.E. (1961). Antibiotics in soil 2. Extent and mechanisms of release. *Soil Science*, **91**, 94–9.

Pirt, S.J (1969). Microbial growth and product formation. In: *Microbial Growth*, 19th Symposium of the Society for General Microbiology, eds P. Meadow & S.J. Pirt, pp. 199–211. Cambridge University Press, Cambridge.

Rothock, C.S. & Gottlieb, G. (1984). Role of antibiosis in antagonism of *Streptomyces hygroscopicus* var *geldans* to *Rhizoctonia solani* in soil. *Canadian Journal of Microbiology*, **30**, 1440–7.

Sanford, G.B. (1926). Some factors affecting the pathogenicity of *Actinomyces scabies*. *Phytopathology*, **16**, 525–47.

Schroth, M.N. & Hancock, J.G. (1981). Selected topics in biological control. *Annual Review of Microbiology*, **35**, 453–76.

Sewell, G.W.F. (1959). Studies of fungi in *Calluna* heathland soil. *Transactions of the British Mycological Society*, **42**, 354–69.

Shipton, P.J., Cook, R.J. & Sihon, J.W. (1973). Occurrence and transfer of a biological factor in soil that suppresses take-all of wheat in Eastern Washington. *Phytopathology*, **63**, 511–17.

Singh, J. & Faull, J.L. (1986). Hyperparasitism and biological control. In: *Biocontrol of Plant Diseases*, eds K.G. Mukerji & K.L. Garg. CRC Press, Boca Raton, Fla.

Smith, J.E. & Moss, M. (1985). *Mycotoxins, Formulation, Analysis and Significance*. Wiley, New York.

Soulides, D.A., Pinck, L.A. & Allison, F.E. (1962). Antibiotics in soil 5, Stability and release of soil absorbed antibiotics. *Soil Science*, **94**, 239–44.

Starr, M.P., Stolp, H., Truper, H.G., Balows, A. & Schlegel, H.G. (1981). *The Prokaryotes, a Handbook on Habitats, Isolation and Identification of Bacteria*,

Vols 1 & 2. Springer-Verlag, Berlin.

Swift, M.J. (1976). Species diversity and the structure of microbial communities in terrestrial habitats. In: *The Role of Terrestrial and Aquatic Organisms in Decomposition Processes*, eds J.M. Andersen & A. Macfadyen, pp. 185–222. Blackwells, Oxford.

Tempest, D.W. & Neijssel, O.M. (1978). Ecophysiological aspects of microbial growth in aerobic, nutrient limited environments. *Advances in Microbial Ecology* 2, ed. M. Alexander, pp. 105–54. Plenum Press, New York.

Warcup, J.H. (1951). The ecology of soil fungi. *Transactions of the British Mycological Society*, **35**, 248–62.

Wicklow, D.T. (1981). Interference competition. In: *The Fungal Community*, eds D.T. Wicklow & G.C. Carroll, pp. 351–78. Marcel Dekker, New York.

Williams, S.T. & Wellington, E.M.H. (1982). Actinomycetes. In: *Methods of Soil Analysis. Part 2. Chemical and Microbiological Properties*, eds A.L. Page, R.H. Millar & D.R. Keeney, pp. 969–87. American Society for Agronomy/Soil Science Society of America, Madison, Wis.

Wright, J.M. (1956a). The production of antibiotics in soil 3. The production of gliotoxin in wheatstraw buried in soil. *Annals of Applied Biology*, **44**, 461–6.

Wright, J.M. (1956b). The production of antibiotics in soil 4. The production of antibiotics in seeds sown in the soil. *Annals of Applied Biology*, **44**, 561–6.

Wright, J.M. (1956c) Biological control of a soilborne *Pythium* infection by seed inoculation. *Plant and Soil*, **8**, 1–9.

Competitive antagonism of air-borne fungal pathogens

8.1 Introduction

Spores of fungal pathogens are deposited on foliar surfaces of plants from the air or by rain splash. Those deposited directly from the air will largely be randomly distributed in relation to surface features of leaves such as veins, trichomes, stomata and lines of junction between adjoining epidermal cells. However, those spores dispersed by water splash often show a distribution pattern relating to the movement and drainage of water over the plant surface. Mostly these spores are situated around the lines of junction between epidermal cells, which act as natural drainage channels from which the spores sediment and become attached by mucopolysaccharides to the leaf. Spores will normally come into contact with microbial antagonists very shortly after arrival on the leaf. Antagonistic organisms, which may be bacteria, yeasts or filamentous fungi, grow on the surface of the leaf at specific favoured sites (Edwards & Blakeman, 1984) which are often similar to those where pathogen spores are deposited. Hence some interaction between the pathogen and saprophytes is inevitable.

The purpose of this review is to examine the nature of such interactions both in relation to the various groups of naturally occurring saprophytic leaf microorganisms and examples of pathogens showing different types of prepenetration and infection behaviour. Understanding of such interactions is an essential prerequisite to implementation of a biological control procedure based on the use of antagonistic microorganisms.

8.2 Microbial succession on foliar surfaces

As antagonists for biological control of plant disease will largely come from naturally occurring saprophytic microorganisms it is necessary to understand some of the factors influencing their growth on leaves at different times during the growing season.

Colonisation of foliar surfaces of temperate plants by saprophytic microorganisms normally follows a fairly distinctive pattern. This is influenced by two main factors: (i) the availability of inoculum of appropriate organisms reaching the surfaces of shoots, and (ii) the provision of favourable environmental conditions at the plant surface for their multiplication. Initially the microflora consists largely of bacteria which live in the very dilute solutions of mainly simple sugars and amino acids leached from internal tissues into water films of rainwater or dew. The bacterial inoculum originates from seed, soil and air (Leben, 1961), and as over-wintering cells on shoots and within buds (Leben, 1971). Bacteria have been shown to have the capacity to migrate from seed or soil and to reach foliar surfaces from these sources provided humidity levels are sufficiently high (Leben, 1961). Early on in the growing season bacteria are often the sole colonists (Warren, 1976; Rodger & Blakeman, 1984). Many bacteria from sycamore leaves were unable to utilise simple sugars such as glucose and were shown to be able to derive both their carbon and nitrogen from amino acids (Rodger, 1981).

As the levels of simple sugars increase on shoot surfaces as a result of deposition of materials such as aphid honeydew, pollen grains and other extraneous organic matter such as fragments of petals, conditions become more favourable for multiplication of yeasts.

Unlike bacteria, leaf-surface yeasts are unable to migrate from seed, soil or older tissues in water films but reach young leaves solely from air-borne inocula (Fokkema *et al.*, 1979). Provision of a suitable nutrient environment together with a delay in arrival of inoculum probably explains why yeast populations do not become established until towards the middle of the growing season. As the number of yeasts increases, bacterial populations tend to decline. This is likely to be due to limitation in availability of amino acids (Blakeman & Brodie, 1977). Yeasts remove large quantities of simple sugars some of which are converted to extracellular polysaccharide. Presence of polysaccharide tends to create an osmotic gradient towards the yeast cell, increasing its ability to absorb solutes (Paton, 1960). This leads to exhaustion of amino acids on the leaf.

Spores of filamentous fungi are deposited from the air on leaves through the growing season but normally fail to establish colonies until the leaves begin to senesce (Dickinson, 1967). At this time relative humidities are higher and nutrients become more available as leakage of solutes from senescing cells increases. Additional nutrient capture may also result from penetration of both living and dying cells (Dickinson, 1981). Later these organisms will be involved in the first stages of saprophytic breakdown of leaves before they fall to the ground.

8.3 Antagonistic capabilities of naturally occurring microorganisms

The choice of a microbial antagonist for use as a control agent against a fungal pathogen will depend on both the nature of its antagonistic properties and the type of inhibitory mechanism to which the pathogen is responsive.

Many species of leaf bacteria can inhibit pathogens by competing for nutrients (Blakeman & Brodie, 1976) whilst fewer are capable of inhibition by direct parasitism (Scherff, 1973) or antibiotic production (Leben & Daft, 1965). Leaf yeasts have been reported to control the development of pathogens solely by competing for nutrients (Fokkema *et al.*, 1979). Certain filamentous fungi have the capability of preventing development of pathogens by direct parasitism, which may be highly specialised, e.g. hyperparasites of fungi (Kranz, 1981), or relatively unspecialised, e.g. parasitism of a wide range of fungi by *Trichoderma* species (Dubos & Bulit, 1981). Antibiotic production is known to be a characteristic of certain leaf-inhabiting filamentous fungi (Andrews, 1985) whilst filamentous fungi are less effective than yeasts or bacteria at competing for nutrients with fungal pathogens on leaves.

8.4 Selection of antagonists for biological control

Antagonists for control of foliar pathogens may be selected either from naturally occurring microorganisms obtained preferably from the surfaces of the host plant it is desired to protect or from 'foreign' organisms of proven antagonistic capabilities but from a different habitat, e.g. soil or roots.

Currently used procedures for selecting antagonists from the phylloplane have been outlined and discussed by Andrews (1985). General methods for isolating microorganisms have been described by Dickinson (1971). It may be considered appropriate initially to screen all organisms isolated from the surfaces of the host for antagonistic activity against the pathogen(s) to be controlled. Alternatively, certain groups of saprophytes may be excluded from tests on the basis that only organisms with certain attributes are likely to be effective against particular types of pathogen. For example it may be reasonable to exclude yeasts from screening tests against rusts and powdery mildews on the grounds that nutrient competitors are unlikely to be effective against obligate pathogens. Instead tests based on appropriate mechanisms of antagonism against this group of pathogens could be set up to screen organisms, e.g. for antibiotic production, lytic enzymes and ability to compete with the pathogen for infection sites, etc.

Both *in vitro* and *in vivo* tests are normally carried out on candidate antagonists. It is now recognised that an organism should not be rejected on the basis of results of *in vitro* tests alone (Andrews, 1985) since failure to inhibit the pathogen on agar or glass slides does not necessarily mean that the organisms will be ineffective on the host plant. Activity against obligate pathogens can, of course, only be tested on the host plant, except for tests on spore germination.

In vivo tests should normally first be carried out using candidate organisms under the controlled conditions of a growth room. If an organism is unable to control the pathogen under constant ideal conditions, it is unlikely to be able to do so in the much more variable environment of the field.

Normally, of the organisms that show activity under growth-room conditions, only a small proportion, if any, would control disease in the field. For success in the field a potential antagonist would need to have the capacity to multiply under conditions favourable to the pathogen and be able to survive in high enough numbers under less favourable conditions in order to rapidly re-exert control on any subsequent return to favourable conditions.

It should also be recognised that in any screening programme not only saprophytic organisms but also pathogens may possess potential for biological control. Suitable candidate organisms in this category might include naturally occurring avirulent strains, pathogens of other hosts not grown in the region or genetically modified pathogens which have been rendered avirulent (see section 8.8).

8.5 Antagonism by fungi

Antagonistic interactions involving yeasts or filamentous fungi with fungal pathogens of shoots of green plants are usually caused by three main mechanisms: hyperparasitism, nutrient competition or antibiotic production.

8.5.1 *Hyperparasitism*

Fungi which parasitise fungal pathogens may be either specialised and relatively specific to their fungal host or largely unspecialised attacking a wide range of fungi. The best known of the latter is probably *Trichoderma viride* which, although not normally associated with foliar surfaces, has been successfully used in this habitat for the control of pathogens, particularly rots of fruits. For example *T. viride* has controlled *Botrytis cinerea* infections on grapes. An inoculum of the antagonist, derived from blended oatmeal agar cultures, was applied four times from the beginning of flowering until three weeks before harvest. The level of control achieved was only slightly inferior to that obtained by use of the fungicide dichlofluanid (Dubos *et al.*, 1978). *Trichoderma harzianum* applied in 0.1% malt extract controlled dry eye rot of apples caused by *B. cinerea* when applied at similar frequencies to fungicides (Tronsmo & Ystaas, 1980). *Botrytis* rot of strawberries was also controlled by *Trichoderma* applications made at time of harvest or during storage (Tronsmo & Dennis, 1977).

Basidiomycetous yeasts of the genus *Tilletiopsis* are commonly found on leaf surfaces. One species, *T. minor*, is a hyperparasite of powdery mildews. In growth-room experiments Hijwegen (1986) found that application of between 10^6 and 2×10^8 cells per millilitre of *T. minor* from 8-day-old cultures to cucumber plants 7 to 9 days after inoculation with cucumber powdery mildew (*Sphaerotheca fuliginea*) effectively reduced mildew infection as determined by the number of healthy conidiophores visible on leaf surfaces. Secondary infections were prevented and plants remained free from powdery mildew infection for a three-week observation period following treatment. *T. minor* was unaffected by dimethirimol fungicide but sensitive to fenarimol, although resistant strains could be obtained after a period of adaptation on media containing fenarimol. *T. minor* could therefore be suitable for use in an integrated programme of control of cucumber powdery mildew.

Previously Sundheim & Amundsen (1982) developed a method of integrated control of cucumber powdery mildew using the hyperparasite *Ampelomyces quisqualis*. The latter was insensitive to the fungicide triforine, which effectively controlled the mildew at reduced rates of application in the presence of the hyperparasite, providing that high humidities were maintained.

Coniothyrium minitans is a well known coelomycete hyperparasite of sclerotia-forming plant pathogens particularly in the soil or seed environment. More recently its effectiveness on aerial plant surfaces has been examined (Trutmann *et al.*, 1982) and it was shown that it failed to control infection by *Sclerotinia sclerotiorum* on bean (*Phaseolus vulgaris*) shoots in the season when it was applied. This was believed to be due to sensitivity to an inhibitor of germination and hyphal growth associated with the surface wax of the leaves. However, by attacking sclerotia associated with crop debris, it reduced the inoculum potential of the pathogen for the succeeding crop. In this respect the authors suggested *C. minitans* would perform better than application of a fungicide such as benlate, although the latter controlled the disease more effectively during the growing season. Integrated control could provide the best method by combining fungicide application to protect the growing crop with post-harvest treatment with *C. minitans* to destroy sclerotia on crop residues.

8.5.2 *Nutrient competition phenomena*
Competition for nutrient reserves leaked from spores of fungal pathogens is largely due to growth of competing bacteria although yeasts can be involved. This is discussed in section 8.6.1 on antagonism by pseudomonads.

The majority of necrotrophic plant pathogens have a temporary saprophytic phase on the plant surface prior to penetration. During this period there is a requirement for exogenous nutrients provided both by nutrients derived from plant leachates and from extraneous materials present on the plant surface. Often the greater the availability of nutrients the more extensive the growth of the pathogen on the plant surface, which can in turn lead to higher infection levels. Yeasts and filamentous fungi, by actively consuming nutrients on the plant surface, are able to limit infection of necrotrophic pathogens such as *Cochliobolus sativus* (Fokkema, 1973).

Pollen grains deposited on leaves provide an excellent nutrient source for stimulating the growth of germ tubes of plant pathogens. Rye plants shed their pollen within a few days and populations of saprophytes are unable to reduce nutrient levels sufficiently rapidly to prevent stimulation of infection of pathogens such as *C. sativus* (Fokkema, 1971). On the other hand the pollen of sugar beet is released gradually over a longer period of time, and growth of the saprophytic microflora prevents nutrients reaching a level at which infection by foliar pathogens is stimulated (Warren, 1972).

Likewise the presence of saprophytes on wheat leaves has been shown to reduce the stimulatory effect of aphid honeydew (Fokkema, 1981).

The capacity of naturally occurring fungal saprophytes to reduce infection by foliar pathogens can be demonstrated after application of fungicides in the presence of fungicide-insensitive strains of pathogens. For example Fokkema *et al.* (1975) showed a reduction of the saprophyte population on rye leaves by a factor of ten after application of benomyl and an increase in infection by a benomyl-insensitive strain of *C. sativus*. This demonstrates the importance of preserving the biological balance between saprophytes and pathogens and shows that fungicides, by upsetting this balance, can lead to increased infection where fungicide-resistant strains have developed.

8.5.3 *Antibiotic production*
Antagonism of foliar pathogens involving nutrient competition mechanisms is widespread among saprophytes on leaves. In contrast there have been relatively few instances where antagonism has been ascribed to antibiotic production.

Antagonism by yeasts is generally thought not to involve antibiotics although a fungistatic substance has been isolated from *Sporobolomyces ruberrimus* (Yamasaki *et al.*, 1951). Although the mycelial yeast-like fungus *Aureobasidium pullulans* has been shown to produce a heat-stable antibiotic in culture (Baigent & Ogawa, 1960), it is likely that the many reports of antagonism of pathogens involving this organism result from nutrient competition.

Of the filamentous fungi associated with foliar plant parts *Botrytis cinerea* is known to produce at least three antimicrobial substances: oxalic acid (Gentile, 1954), 'Botrycine', a high molecular weight polysaccharide which inhibits yeasts in grape musts (Ribereau-Gayon *et al.*, 1979), and 'Botrydial', a bicyclic non-isoprenoid sesquiterpene active against fungi and Gram-positive bacteria. Strains of *Alternaria* also produce substances which are active against fungi (Lindenfelser and Ciegler, 1969). *Trichoderma* spp., although not strictly a phylloplane genus, can often be isolated from leaves especially after heavy rain when spores are splashed on to plant foliage from soil. Both non-volatile (Dennis & Webster, 1971a) and volatile (Dennis & Webster, 1971b) antibiotics are produced.

During a comprehensive screening programme of microorganisms from apple leaves for activity against the scab pathogen, *Venturia inaequalis*, Andrews *et al.* (1983) found that one of the most effective antagonists was *Chaetomium globosum*. Evidence suggested that the action on *V. inaequalis* of *C. globosum* was associated with antibiotic production.

Trichothecium roseum has been shown to actively antagonise the pathogen *Pestalotia funerea* on cones of *Thuja occidentalis*. The antagonist was shown *in vitro* to produce water-soluble, heat-stable substances which inhibited germination of conidia and hyphal growth of *P. funerea* (Urbasch, 1985).

8.6 Antagonism by bacteria

The majority of antagonistic interactions which have been described between bacteria and fungal pathogens on foliar surfaces have involved bacteria from two genera, *Pseudomonas* and *Bacillus*.

8.6.1 *Pseudomonads*

Although many bacteria and yeasts can inhibit germination of spores of fungal pathogens on leaves, populations of pseudomonads can often predominate and have been shown to compete actively for nutrients (Blakeman, 1972; Blakeman & Brodie, 1977).

Leakage of solutes from spores and germ tubes of pathogens provides an additional source of nutrients for bacteria on plant surfaces (Fraser, 1971). This has been shown to result in enhanced populations of bacteria in the immediate vicinity of fungal spores. The observed inhibition of germination has been considered similiar to fungistatic effects reported for fungal spores in soil (Blakeman, 1978). The majority of nutrients leaked from *Botrytis cinerea* conidia occurred within a few minutes of suspending in water (Brodie & Blakeman, 1975). When bacteria were present around conidia, the leaked nutrients were preferentially absorbed by the bacteria, making them unavailable to the conidia. Consequently germination was inhibited. In the absence of bacteria, leaked nutrients were reabsorbed by conidia, enhancing germination (Brodie & Blakeman, 1977). In nature, loss of endogenous nutrients from fungal conidia may be of less significance because nutrients are also supplied exogenously from the plant surface. However, competition from bacteria can also greatly restrict the supply of nutrients to germinating conidia (Table 8.1). Of a number of different bacteria examined from red beet (*Beta vulgaris* L.) leaves, species of *Pseudomonas* were found to be the most active both in competing for amino acids and in inhibiting germination of *Botrytis cinerea* and other necrotrophic pathogens such as *Phoma betae* and *Cladosporium herbarum* (Blakeman & Brodie, 1977).

Certain members of the genus *Pseudomonas* are known to produce antibiotics which are effective against foliar pathogens. Isolates of *Pseudomonas* and other bacteria obtained from lettuce leaves infected with *B. cinerea* produced inhibition zones against the pathogen on agar and gave some control of the disease when inoculated on to plants (Newhook, 1951). *Ps. fluorescens* is widely distributed as an epiphyte on root and shoot surfaces and has been shown to produce two antifungal antibiotics (Howell

Table 8.1. Numbers of natural bacterial epiphytes, uptake of amino acids and germination of *Botrytis cinerea* conidia on unwetted and previously wetted beetroot leaves (data from Blakeman & Brodie, 1977)

	No. of bacteria ($\times 10^6$)	Uptake of amino acids (%) after 24h	Germination (%) of Botrytis cinerea conidia after 24h
Previously unwetted leaves	0.6	2.1	80
Leaves previously wetted for 24h	3.7	76.4	17

& Stipanovic, 1980). *Ps. cepacia*, isolated from corn, produced inhibition zones against *Drechslera maydis* on agar (Sleesman & Leben, 1976). The same organism was subsequently used to control *Cercospora* leaf spot on peanut and *Alternaria* leaf spot on tobacco (Spurr, 1981).

Unlike certain *Bacillus* species discussed below there appear to be no reports of pseudomonads directly parasitising fungi on aerial plant surfaces.

8.6.2 *Bacilli*

Species of *Bacillus* have been reported to affect fungal pathogens of aerial plant parts by two main mechanisms, parasitism and antibiotic production.

Bacillus pumilus markedly reduced infection by the rust *Puccinia recondita* when applied to cereal leaves (Morgan, 1963). A similar effect was observed even if the bacterium was applied under controlled conditions up to 9 days before inoculation of the rust, indicating some ability of the bacterium to persist on leaves. *B. pumilus* was shown to lyse uredospore germ tubes by the production of a heat-stable lytic substance. Other isolates of *Bacillus* spp. had a similar effect on the rust. It is not only uredospore germ tubes that are lysed by species of *Bacillus* but also fruiting structures of rusts including pycnia, aecidia and uredia (Levine *et al.*, 1936).

Many different collections of uredospores of *Puccinia allii* obtained from various parts of the United Kingdom were examined by Doherty & Preece (1978) for the presence of *B. cereus*. The bacterium was readily isolated from most of the samples received. Application of *B. cereus* to the leek host leaves greatly reduced the number of rust pustules formed but this effect could not be reproduced by culture filtrates of the bacterium despite the fact that the effect on uredospores produced by living cells of the bacterium could pass through cellophane. Possibly an unstable or volatile inhibitory factor may have been involved because uredospores on agar could be inhibited at a distance from areas of the medium inoculated with the bacterium.

A number of isolates of bacteria which included *Bacillus subtilis*, *B. cereus* subsp. *mycoides* and *B. thuringiensis* were able to control development of bean rust, *Uromyces phaseoli*, on snap and dry beans (*Phaseolus vulgaris* L.) (Baker *et al.*, 1983). An isolate of *B. subtilis* was found to have the greatest effect, reducing rust pustule numbers by over 95% when sprayed on to plants in a greenhouse from between 2 to 120 h before inoculation with uredospores. The bacterium reduced uredospore germination, prevented normal germ-tube development and caused cytoplasmic abnormalities in the uredospores. Because of these effects on germination it is therefore not surprising that no control of the disease was achieved when bacterial cells were applied after uredospore inoculation. A largely protein-containing heat-stable non-dialysable component was detected in culture filtrates which both inhibited uredospore germination and reduced bean rust.

Later field tests showed that two isolates of *B. subtilis* could reduce bean rust by over 75% (Baker *et al.*, 1985). As a result of applying *B. subtilis* three times a week the level of control of bean rust was similar to that obtained with one application per week of the fungicide mancozeb. One of the isolates of *B. subtilis* tested appeared to have an effect on promoting vegetative growth of bean at the expense of flowering and fruiting. This would clearly be unsuitable for biological control purposes.

Antibiotics have been implicated in the majority of antagonistic interactions involving *B. subtilis* and plant pathogens. McKeen *et al.* (1986) extracted four cyclic polypeptides from *B. subtilis* culture filtrates by precipitation after acidification and extraction with ethanol. The compounds were active against a wide range of plant pathogenic fungi. Loeffler *et al.* (1986) detected two antibiotics from *B. subtilis*. One, a dipeptide, bacilysin, inhibited yeasts and bacteria, whereas fengymycin, a complex of lipopeptides, was active against filamentous fungi. Fengymycin proved to be more active than bacilysin against certain pathogens such as *Rhizoctonia solani* on rice and it was less phytotoxic. Bacilysin was produced by many strains of *B. subtilis* and also by some other closely related species in the genus.

Bacillus subtilis has been shown to be naturally present in Northern Ireland on apple (cv. Bramley's Seedling) leaf scar tissue, which is a site for entry and infection by the apple canker pathogen, *Nectria galligena* (Swinburne, 1973). The bacterium can persist on the scar tissue over winter and reduce incidence of canker the following spring. The bacterium was found to produce two antifungal antibiotics causing swelling and bursting of *N. galligena* germ tubes. On agar one of the antibiotics was relatively stable and produced inhibition zones against *N. galligena* (Swinburne *et al.*,

1975). However, attempts to increase natural populations of *B. subtilis* on apple leaf scar tissue failed because leaf fall in cv. Bramley's Seedling occurs over an extended period. This meant that a single spray with *B. subtilis* would only enable the bacterium to reach a small proportion of leaf scars, the bacterium having no ability to survive on bark and recolonise scars subsequently after leaves had fallen. Of a number of bacteria which included *Pseudomonas cepacia*, *P. fluorescens*, *Bacillus thuringiensis* and two isolates of *B. subtilis* applied to peaches, nectarines, apricots and plums to control brown fruit rot caused by *Monilinia fructicola*, only an isolate of *B. subtilis* controlled brown rot on all fruits (Pusey and Wilson, 1984). Most effective control occurred on peaches and apricots and was apparent at all temperatures at which brown rot developed. The activity of *B. subtilis* was not affected in the presence of the fungicide dichloran used for control of *Rhizopus* and was only slightly reduced by commercial fruit waxes (Pusey *et al.*, 1986). As with bean rust discussed earlier, the bacterium acted by inhibiting spore germination and early germ-tube development with little effect on subsequent hyphal growth. The causal agent produced by the bacterium was believed to be a low molecular weight polar substance which could reproduce the effects of the bacterium on *M. fructicola* spores *in vitro*. Brown rot infection of apple fruits was also reduced by *B. cereus* (Jenkins, 1968). Similarly the effect of the bacterium was on spore germination and could also be reproduced by culture filtrates.

Bacillus cereus var. *mycoides* (Flügge) Smith was found to be a common epiphyte on Douglas fir needles where, when populations were increased by applying the cells of the bacterium in a nutrient broth medium, control of the rust *Melampsora medusae* Thum was achieved under greenhouse conditions (McBride, 1969). Antibiotic production was a likely cause of this effect since some control of the rust could also be achieved with cell-free culture filtrates of the bacterium.

8.7 Behaviour of different groups of fungal pathogens in relation to antagonists

It is possible to predict the response of individal pathogens to different mechanisms of antagonism according to broad ecological groupings in which plant pathogens can be categorised (Table 8.2). For example, the response of biotrophic pathogens to antagonists may be very different from that of necrotrophic pathogens.

8.7.1 *Unspecialised necrotrophs*
Necrotrophic foliar pathogens comprise relatively unspecialised fungi such as *B. cinerea* and species of *Alternaria*, *Cladosporium*, *Cochliobolus* and *Septoria*. Prior to penetration they characteristically grow saprophytically

Table 8.2.   Sensitivity of groups of pathogens to different antagonistic mechanisms associated with saprophytic microorganisms on foliar surfaces

Pathogen group	Mechanism sensitivity	Appropriate antagonist group
Biotrophic	Antibiotics	Bacteria, filamentous fungi
	Parasitism	Filamentous fungi, yeasts
Specialised necrotrophic	Nutrient competition	Bacteria, yeasts
Unspecialised necrotrophic	Antibiotics ⎫ Parasitism ⎬	Bacteria, filamentous fungi

on the plant surface, deriving nutrients both from extraneous deposits and from leachates from the leaf. Pathogens in this group are therefore most exposed to competition at this stage and, because of their requirement for nutrients, they widely respond to saprophytes which are active nutrient competitors. This mechanism of antagonism has been discussed previously in section 8.5.2. Necrotrophic pathogens can also be inhibited by saprophytic organisms which can bring about germ-tube lysis or produce antibiotics.

8.7.2 *Anthracnoses*

Antagonists may influence the behaviour of anthracnose fungi on the surfaces of leaves and fruits in a variety of different ways. Conidia and germ tubes of *Colletotrichum gloeosporioides* Penz. have been observed to be lysed on host leaves by groups of bacteria (Lenné & Parbery, 1976). However, if appressoria formed before lysis occurred, these were unaffected even if spores and germ tubes were subsequently lysed by bacteria. In *C. musae* there is strong evidence to suggest that appressoria function as survival structures (Muirhead, 1981). Previous evidence which indicated that subcuticular hyphae rather than dark appressoria were the dormant structures was questioned when it was shown that appressoria survived treatment with mercuric chloride and subsequently germinated. Hyaline appressoria which were not dormant germinated directly to form subcuticular hyphae which ceased growth permanently when they induced a hypersensitive response in green, unripe fruit. It is possible that a similar mechanism may be responsible for latency in other anthracnoses.

For example, when apparently disease-free leaf and stem samples of the tropical legume stylo (*Stylosanthes guianensis*) collected from pastures in various sites in the humid tropics of South America were incubated under humid conditions in the laboratory, typical *C. gloeosporioides* infections developed (Lenné, 1986).

Additional observations (Blakeman & Brodie, 1977) showed that although the proportion of germinating conidia of *Colletotrichum dematium* f. sp. *spinaceae* on beetroot leaves was not affected in the presence of bacteria, growth of germ tubes was markedly reduced. Appressoria were formed earlier, either on short germ tubes or in a sessile form immediately adjacent to the spores. The response of *Colletotrichum* conidia to the presence of bacteria on leaves thus differed from unspecialised necrotrophic pathogens such as *Botrytis*, *Phoma* and *Cladosporium* where the proportion of germinating conidia was markedly reduced. Further studies with *Colletotrichum acutatum* (Blakeman & Parbery, 1977) showed similar effects. When tested with individual isolates of leaf bacteria and yeasts, it was found that those that competed most actively for amino acids caused the greatest increase in numbers of appressoria, indicating that competition for nutrients was likely to be a causal factor. Further evidence for nutrient competition was shown in leaching experiments on membranes where more appressoria were induced when spores were leached with water than occurred when leached with an amino acid/glucose solution or in unleached static controls. Previously Mercer *et al.* (1970) reported that nutrient solutions such as orange juice increased extension growth of hyphae of *Colletotrichum lindemuthianum* on French bean leaves but depressed formation of appressoria.

Conidia of anthracnose fungi are sensitive to the presence of iron in their environment. For example, germination of *Colletotrichum musae* (Berk & Curt) v. Arx is inhibited in the presence of iron-containing salts but stimulated in the presence of a naturally occurring iron-chelating compound that was found to be present in banana fruit leachates (Swinburne, 1976). The active component in the leachates was shown to be anthranilic acid, a weak chelating agent, which was converted by *C. musae* to 2,3-dihydroxybenzoic acid, a compound possessing greater iron-chelating activity (Harper & Swinburne, 1979). Another source of iron chelators was the saprophytic bacteria growing on banana fruit surfaces which produced compounds known as siderophores which stimulated germination and also caused some increase in numbers of appressoria formed (McCracken & Swinburne, 1979). The findings reported above were supported by the observation that conidia produced on iron-deficient culture media germinated well in the absence of chelating agents.

From the evidence presented above it would appear that leaf microorganisms, especially bacteria present in the vicinity of germinating *Colletotrichum* conidia, might increase infection as a result of stimulating germination and appressorium formation. Is there any way in which nutrient competing microorganisms can reduce infection by anthracnose pathogens? Recently Williamson & Fokkema (1985) have shown that leaf yeasts such as species of *Sporobolomyces* and *Cryptococcus* can reduce infection by 50% of *Colletotrichum gramincola* on maize plants. Both the number of lesions and the degree of necrosis were affected. The presence of the yeasts did not affect spore germination or appressoria formation, but the authors identified the site of antagonism as the germination of the appressorium to form the infection peg. The authors suggest that the yeast cells acted as a nutrient sink reducing the availability of nutrients to the germinating appressoria. This would indicate that yeasts were either removing endogenous nutrients from the appressoria or preventing access to exogenous nutrients during the penetration process. There was no visual indication that the yeasts reduced melanin biosynthesis by the appressoria which might have reduced penetration since studies by Wolkow *et al.* (1983) have shown that certain chemicals that inhibit melanin biosynthesis prevent penetration. It is known that cutinase inhibitors prevent infection by anthracnose fungi (Kolattukudy, 1984) and it is possible that yeasts directly interfere with synthesis of this enzyme or reduce nutrients required for its synthesis. It is thus unclear exactly what the precise mechanism is which enables competing yeasts to interfere with penetration by appressoria.

8.7.3 *Biotrophs*

Although biotrophic pathogens, such as rusts and powdery mildews, may produce fairly long germ tubes from spores prior to penetration of the leaf, they exhibit no saprophytic activity during this phase. Using C^{14} labelling, studies have shown that almost no exogenous substrates are consumed by rust uredospores compared with saprophytes prior to entry of the plant (Staples *et al.*, 1962). This would explain why antagonists which are solely nutrient competitors are ineffective at inhibiting the germination and germ tube growth of biotrophic pathogens. On the other hand such pathogens can be antagonised by organisms producing lytic enzymes or antibiotics or can be attacked by hyperparasites as discussed in earlier sections.

Saprophytic organisms, especially yeasts and the yeast-like fungus *Aureobasidium pullulans*, can enhance germination and growth of germ tubes of the rusts *Uromyces viciae-fabae* (Parker & Blakeman, 1984) and *Puccinia antirrhini* (Collins, 1974). Experiments with *U. viciae-fabae* showed that yeasts of the genus *Cryptococcus* were most stimulatory and the effect could be reproduced with cell-free leachates even at high dilutions. Application of yeast cells to broad bean (*Vicia faba*) leaves, either simultaneously with the rust or 2 to 3 days before, resulted in a substantial increase in numbers of rust pustules (Parker & Blakeman, 1984).

8.8 Genetically modified pathogens as competitors

A plant pathogen which has been genetically modified to render it non-pathogenic may act as a very effective competitor against the unmodified pathogen from which it was derived. Nutrient requirements would be similar to those of the wild type as would be the infection sites for colonisation of plant surfaces. Application of the genetically modified organism and allowing it to colonise the plant surface could therefore result in reduction in infection by the pathogen by a process of exclusion. There have been a number of attempts to modify bacterial pathogens but as yet work of this kind has not been attempted with fungal pathogens. As with bacteria (Lindow, 1986), fungal pathogens could be modified by deletion of one or more genes affecting virulence. Other attributes associated with pathogenicity might also be considered for deletion such as genes controlling phytotoxin production or those associated with the formation of infection structures, hormone production in the case of pathogens that induce hypertrophy and hyperplasia of host tissues or production of macerating enzymes in many necrotrophic pathogens.

8.9 Conclusions

A large number of different microorganisms, either naturally occurring on shoot surfaces or obtained from other habitats, have been shown to possess the potential to limit the growth of foliar plant pathogens *in vitro* or on the host plant under controlled or semi-controlled environmental conditions. However, very few candidate antagonists are capable of controlling plant disease in the field as effectively as modern fungicides. Even the most effective organisms need to be applied to the crop frequently — usually at least as often as recommended application times for fungicides. The potential advantage of using an organism, as opposed to a chemical, to control disease is that the organism should be able to multiply and maintain itself on the plant thereby providing continuous protection over long periods of time. The fact that this potentially desirable objective has rarely been achieved is probably due to a lack of understanding of the character-istics of candidate antagonists and the nature of the foliar environment. An antagonist must be selected not only for its activity against the pathogen but also for its capacity to survive adverse environmental conditions and ability to maintain itself at high enough population levels to be effective.

It is likely that comparatively few organisms will meet such demanding requirements, particularly as shoot surfaces face greater extremes of physical environmental conditions than below-ground parts of plants. It may well be that genetic engineering techniques will be used to introduce characters that confer a high level of antagonism into an organism of proven ability to survive and maintain itself on the host in question. Another possibility is that the pathogen itself could be modified to render it avirulent. It may then effectively exclude wild-type virulent strains by taking up infection sites or inducing defence reactions by the host.

References

Andrews, J.H. (1985). Strategies for selecting antagonistic microorganisms from the phylloplane. In: *Biological Control on the Phylloplane*, eds C.E. Windels & S.E. Lindow, pp. 31–44. American phytopathological Society, St. Paul, Minn.
Andrews, J.H., Berbee, F.M. & Nordheim, E.V. (1983). Microbial antagonism to the imperfect stage of the apple scab pathogen, *Venturia inaequalis. Phytopa-thology*, **73**, 228–34.
Baigent, N.L. & Ogawa, J.M. (1960). Activity of the antibiotic produced by *Pullularia pullulans. Phytopathology*, **50**, 82 (Abstr.).
Baker, C.J., Stavely, J.R. & Mock, N. (1985). Biocontrol of bean rust by *Bacillus*

subtilis under field conditions. *Plant Disease*, **69**, 770–2.

Baker, C.J., Stavely, J.R., Thomas, C.A., Sasser, M. & MacFall, J.S. (1983). Inhibitory effect of *Bacillus subtilis* on *Uromyces phaseoli* and on development of rust pustules on bean leaves. *Phytopathology*, **73**, 1148–52.

Blakeman, J.P. (1972). Effect of plant age on inhibition of *Botrytis cinerea* spores by bacteria on beetroot leaves. *Physiological Plant Pathology*, **2**, 143–52.

Blakeman, J.P. (1978). Microbial competition for nutrients and germination of fungal spores. *Annals of Applied Biology*, **89**, 151–55.

Blakeman, J.P. & Brodie, I.D.S. (1976). Inhibition of pathogens by epiphytic bacteria on aerial plant surfaces. In: *Microbiology of Aerial Plant Surfaces*, eds C.H. Dickinson & T.F. Preece, pp. 529–57. Academic Press, London.

Blakeman, J.P. & Brodie, I.D.S. (1977). Competition for nutrients between epiphytic micro-organisms and germination of spores of plant pathogens on beetroot leaves. *Physiological Plant Pathology*, **10**, 29–42.

Blakeman, J.P. & Parbery, D.G. (1977). Stimulation of appressorium formation in *Colletotrichum acutatum* by phylloplane bacteria. *Physiological Plant Pathology*, **11**, 313–25.

Brodie, I.D.S. & Blakeman, J.P. (1975). competition for carbon compounds by a leaf surface bacterium and conidia of *Botrytis cinerea*. *Physiological Plant pathology*, **6**, 125–35.

Brodie, I.D.S. & Blakeman, J.P. (1977). Effect of nutrient leakage, respiration and germination of *Botrytis cinerea* conidia caused by leaching with water. *Transactions of the British Mycological Society*, **68**, 445–7.

Collins, M.A. (1974). Studies on some leaf surface microorganisms with special reference to the phylloplane of *Antirrhinum majus* L. PhD thesis, University of Edinburgh, 296 pp.

Dennis, C. & Webster, J. (1971a). Antagonistic properties of species groups of *Trichoderma* I. Production of non-volatile antibiotics. *Transactions of the British Mycological Society*, **57**, 25–39.

Dennis, C. & Webster, J. (1971b). Antagonistic properties of species groups of *Trichoderma* II. Production of volatile antibiotics. *Transactions of the British Mycological Society*, **57**, 41–8.

Dickinson, C.H. (1967). Fungal colonisation of *Pisum* leaves. *Canadian Journal of Botany*, **45**, 915–27.

Dickinson, C.H. (1971). Cultural studies of leaf saprophytes. In: *Ecology of Leaf Surface Micro-organisms*, eds T.F. Preece & C.H. Dickinson, pp. 129–37. Academic Press, London.

Dickinson, C.H. (1981). Biology of *Alternaria alternata*, *Cladosporium cladosporioides* and *C. herbarum* in respect of their activity on green plants. In: *Microbial Ecology of the Phylloplane*, ed. J.P. Blakeman, pp. 169–84. Academic Press, London.

Doherty, M.A. & Preece, T.F. (1978). *Bacillus cereus* prevents germination of uredospores of *Puccinia allii* and the development of rust disease of leek, *Allium porrum*, in controlled environments. *Physiological Plant Pathology*, **12**, 123–32.

Dubos, B. & Bulit, J. (1981). Filamentous fungi as biocontrol agents on aerial plant surfaces. In: *Microbial Ecology of the Phylloplane*, ed. J.P. Blakeman, pp. 353–6. Academic Press, London.

Dubos, B., Bulit, J., Bugaret, Y. & Verdu, D. (1978). Possibilités d'utilisation de *Trichoderma viride* Pers. comme moyen biologique de lutte contre la pourriture grise (*Botrytis cinerea* Pers.) et l'excoriose (*Phomopsis viticola* Sacc.) de la vigne. *Comptes Rendus Academie Agriculture Français*, **64**, 1159–68.

Edwards, M.C. & Blakeman, J.P. (1984). An autoradiographic method for determining nutrient competition between leaf surface epiphytes and plant pathogens. *Journal of Microscopy*, **133**, 205–12.

Fokkema, N.J. (1971). The effect of pollen in the phyllosphere of rye on colonisation by saprophytic fungi and on infection by *Helminthosporium sativum* and other leaf pathogens. *Netherlands Journal of Plant Pathology*, **77**, Suppl. 1, pp. 1–60.

Fokkema, N.J. (1973). The role of saprophytic fungi in antagonism against *Drechslera sorokiniana* (*Helminthosporium sativum*) on agar plates and on rye leaves with pollen. *Physiological Plant Pathology*, **3**, 195–205.

Fokkema, N.J. (1981). Fungal leaf saprophytes, beneficial or detrimental? In: *Microbial Ecology of the Phylloplane*, ed. J.P. Blakeman, pp. 433–54. Academic Press, London.

Fokkema, N.J., den Houter, J.G., Kosterman, Y.J.C. & Nelis, A.L. (1979). Manipulation of yeasts on field-grown wheat leaves and their antagonistic effect on *Cochliobolus sativus* and *Septoria nodorum*. *Transactions of the British Mycological Society*, **72**, 19–29.

Fokkema, N.J., van de Laar, J.A.J., Nelis-Blomberg, A.L. & Schippers, B. (1975). The buffering capacity of the natural mycoflora of rye leaves to infection by *Cochliobolus sativus* and its susceptibility to benomyl. *Netherlands Journal of Plant Pathology*, **81**, 176–86.

Fraser, A.K. (1971). Growth restriction of pathogenic fungi on the leaf surface. In: *Ecology of Leaf Surface Micro-organisms*, eds T.F. Preece & C.H. Dickinson, pp. 529–35. Academic Press, London.

Gentile, A.C. (1954). Carbohydrate metabolism and oxalic acid synthesis by *Botrytis cinerea*. *Plant Physiology*, **29**, 257–61.

Harper, D.B. & Swinburne, T.R. (1979). 2-3-Dihydroxy-benzoic acid and related compounds as stimulants of germination of conidia of *Colletotrichum musae* (Berk. & Curt.) Arx. *Physiological Plant Pathology*, **14**, 363–70.

Hijwegen, T. (1986). Biological control of cucumber powdery mildew by *Tilletiopsis minor*. *Netherlands Journal of Plant Pathology*, **92**, 93–5.

Howell, C.R. & Stipanovic, R.D. (1980). Suppression of *Pythium ultimum-induced* damping off of cotton seedlings by *Pseudomonas fluorescens* and its antibiotic, pyoluteorin. *Phytopathology*, **70**, 712–15.

Jenkins, P.T. (1968). The longevity of conidia of *Sclerotinia fructicola* (Wint.) Rehm under field conditions. *Australian Journal of Biological Science*, **21**, 937–45.

Kolattukudy, P.E. (1984). Fungal penetration of defensive barriers of plants. In: *Structure, Function and Biosynthesis of Plant Cell Walls*, eds W.M. Duggar & S. Bartnicki-Garcia, pp. 302–43. *Proceedings of the Seventh Symposium of Botany*, University of California, Riverside.

Kranz, J. (1981). Hyperparasitism of biotrophic fungi. In: *Microbial Ecology of the Phylloplane*, ed. J.P. Blakeman, pp. 327–52. Academic press, London.

Leben, C. (1961). Microorganisms on cucumber seedlings. *Phytopathology*, **51**, 553–7.

Leben, C. (1971). The bud in relation to the epiphytic microflora. In: *Ecology of Leaf Surface Micro-organisms*, eds T.F. Preece & C.H. Dickinson, pp. 117–27. Academic Press, London.

Leben, C. & Daft, G.C. (1965). Influence of an epiphytic bacterium on cucumber anthracnose, early blight of tomato and northern leaf blight of corn. *Phytopathology*, **55**, 760–2.

Lenné, J.M. (1986). Recent advances in the understanding of anthracnose of

Stylosanthes in tropical America. In: *Proceedings of the XVth International Grasslands Congress, 24–31 August, Kyoto, Japan*, 773–5.

Lenné, J.M. & Parbery, D.G. (1976). Phyllosphere antagonists and appressorium formation in *Colletotrichum gloeosporioides. Transactions of the British Mycological Society*, **66**, 334–6.

Levine, M.N., Bamberg, R.H. & Atkinson, R.E. (1936). Micro-organisms antibiotic or pathogenic to cereal rusts. *Phytopathology*, **26**, 99–100.

Lindenfelser, L.A. & Ciegler, A. (1969). Production of antibiotics by *Alternaria* species. *Developments in Indian Microbiology*, **10**, 271–8.

Lindow, S.E. (1986). *In vitro* construction of biological control agents. In: *Biotechnology and Crop Improvement and Protection, BCPC Monograph No. 34*, 185–98.

Loeffler, W., Tschen, S.M., Vanittanakom, N., Kugler, M., Knorpp, E., Hsieh, T.F. & Wu, T.G. (1986). Antifungal effects of bacilysin and fengymycin from *Bacilus subtilis* F-29-3. A comparison with activities of other *Bacillus* antibiotics. *Journal of Phytopathology*, **115**, 204–13.

McBride, R.P. (1969). A microbiological control of *Melampsora medusae. Canadian Journal of Botany*, **47**, 711–15.

McCracken, A.R. & Swinburne, T.R. (1979). Siderophores produced by saprophytic bacteria as stimulants of germination of conidia of *Colletotrichum musae. Physiological Plant Pathology*, **15**, 331–40.

McKeen, C.D., Reilly, C.C. & Pusey, P.L. (1986). Production and partial characterisation of antifungal substances antagonistic to *Monilinia fructicola* from *Bacillus subtilis. Phytopathology*, **76**, 136–9.

Mercer, P.C., Wood, R.K.S. & Greenwood, A.D. (1970). The effect of orange extract and other additives on anthracnose of French beans caused by *Colletotrichum lindemuthianum. Annals of Botany*, **34**, 593–604.

Morgan, F.L. (1963). Infection inhibition and germ tube lysis of three cereal rusts by *Bacillus pumilus. Phytopathology*, **53**, 1346–8.

Muirhead, I.F. (1981). The role of appressorial dormancy in latent infection. In: *Microbial Ecology of the Phylloplane*, ed. J.P. Blakeman, pp. 155–67. Academic Press, London.

Newhook, F.J. (1951). Microbiological control of *Botrytis cinerea* Pers. I. The role of pH changes and bacterial antagonism. *Annals of Applied Biology*, **38**, 169–84.

Parker, A. & Blakeman, J.P. (1984). Stimulation of *Uromyces viciae-fabae in vitro* and *in vivo* by the phylloplane yeast *Cryptococcus. Physiological Plant Pathology*, **24**, 119–28.

Paton, A.M. (1960). The role of *Pseudomonas* in plant disease. *Journal of Applied Bacteriology*, **23**, 526–32.

Pusey, P.L. & Wilson, C.L. (1984). Postharvest biological control of stone fruit brown rot by *Bacillus subtilis. Plant Disease*, **68**, 753–6.

Pusey, P.L., Wilson, C.L., Hotchkiss, M.W. & Franklin, J.D. (1986). Compatibility of *Bacillus subtilis* for postharvest control of peach brown rot with commercial fruit waxes, dicloran and cold-storage conditions. *Plant Disease*, **70**, 587–90.

Ribéreau-Gayon, P., LaFon-LaFourcade, S., Dubourdieu, D., Lucmaret, V. & Larue, F. (1979). Metabolisme de *Saccharomyces cerevisiae* dans le moit de raisins parasites par *Botrytis cinerea*. Inhibition de la fermentation; formation d'acide acetique et de glycerol. *Comptes Rendus Academic des Sciences Paris D*, **289**, 441–4.

Rodger, G. (1981). Microbial competition for nutrients on leaves of sycamore and lime. PhD thesis, University of Aberdeen, 101 pp.

Rodger, G. & Blakeman, J.P. (1984). Microbial colonization and uptake of ^{14}C label on leaves of sycamore. *Transactions of the British Mycological Society*, **82**, 45–51.

Scherff, R.H. (1973). Control of bacterial blight of soybean by *Bdellovibrio bacteriovorus*. *Phytopathology*, **63**, 400–2.

Sleesman, J.P. & Leben, C. (1976). Microbial antagonists of *Bipolaris maydis*. *Phytopathology*, **66**, 1214–18.

Spurr, H.W. (1981). Experiments on foliar disease control using bacterial antagonists. In: *Microbiol Ecology of the Phylloplane*, ed. J.P. Blakeman, pp. 369–81. Academic Press, London.

Staples, R.C., Syamananda, R., Kao, V. & Block, R.J. (1962). Comparative biochemistry of obligately parasitic and saprophytic fungi. II. Assimilation of C^{14}-labelled substrates by germinating spores. *Contributions of the Boyce Thompson Institute*, **21**, 345–62.

Sundheim, L. & Amundsen, T. (1982). Fungicide tolerance in the hyperparasite *Ampelomyces quisqualis* and integrated control of cucumber powdery mildew. *Acta Agriculturae Scandinavica*, **32**, 349–55.

Swinburne, T.R. (1973). Microflora of apple leaf scars in relation to infection by *Nectria galligena*. *Transactions of the British Mycological Society*, **60**, 389–403.

Swinburne, T.R. (1976). Stimulants of germination and appressoria formation by *Colletotrichum musae* (Berk. & Curt.) Arx in banana leachate. *Phytopathologische Zeitschrift*, **87**, 74–90.

Swinburne, T.R., Barr, J. & Brown, A.E. (1975). Production of antibiotics by *Bacillus subtilis* and their effect on fungal colonists of apple leaf scars. *Transactions of the British Mycological Society*, **65**, 211–17.

Tronsmo, A. & Dennis, C. (1977). The use of *Trichoderma* species to control strawberry fruit rots. *Netherlands Journal of Plant Pathology*, **83**, Suppl. 1, 449–5

Tronsmo, A. & Ystaas, J. (1980). Biological control of *Botrytis cinerea* on apple. *Plant Disease*, **64**, 1009.

Trutmann, P., Keane, P.J. & Merriman, P.R. (1982). Biological control of *Sclerotina sclerotiorum* on aerial parts of plants by the hyperparasite *Coniothyrium minitans*. *Transactions of the British Mycological Society*, **78**, 521–9.

Urbasch, I. (1985). Antagonistische Wirkung von *Trichothecium roseum* (Pers.) Link ex Gray auf *Pestalotia funerea* Desm. *Phytopathologische Zeitschrift*, **113**, 343–7.

Warren, R.C. (1972). The effect of pollen on the fungal leaf microflora of *Beta vulgaris* L. and on infection of leaves by *Phoma betae*. *Netherlands Journal of Plant Pathology*, **78**, 89–98.

Warren, R.C. (1976). Occurrence of microbes among buds of deciduous trees. *European Journal of Forest Pathology*, **6**, 38–45.

Williamson, M.A. & Fokkema, N.J. (1985). Phyllosphere yeasts antagonize penetration from appressoria and subsequent infection of maize leaves by *Colletotrichum graminicola*. *Netherlands Journal of Plant Pathology*, **91**, 265–76.

Wolkow, P.M., Sisler, H.D. & Vigil, E.L. (1983). Effect of inhibitors of melanin biosynthesis on structure and function of appressoria of *Colletotrichum lindemuthianum*. *Physiological Plant Pathology*, **23**, 55–71.

Yamasaki, I., Satomura, Y. & Yamamoto, T. (1951). Studies on *Sporobolomyces* red yeast. 10. Antidiabetic action and fungistatic action (Part 3). *Journal of the Agricultural Society of Japan*, **24**, 399–402.

9 *J.M. Whipps, Karen Lewis and R.C. Cooke*

Mycoparasitism and plant disease control

9.1 Introduction

A number of important plant diseases are susceptible to control by mycoparasites yet in commercial terms their potential remains largely unrealised.
The reasons for this are complex, but major factors are a lack of detailed
information on mycoparasitic behaviour in nature, as opposed to that in
controlled conditions, and consequent failure to generate strong principles
upon which biocontrol studies might be based. All too frequently an
empirical approach has been used, with efficacy of biological control systems being assessed without first obtaining essential data concerning either
mycoparasites or the habitats into which they were to be introduced. This
has contributed to a general inability to quantify accurately the possible
environmental and economic benefits of mycoparasite-based biological

control systems in relation to those achieved through more conventional disease management methods.

What follows is an attempt to take a selective and fundamental eco-physiological rather than phytopathological view of mycoparasitism as a phenomenon within the context of the agricultural ecosystems of which mycoparasites are, or may be induced to become, important components.

9.2 Mycoparasitism

A mycoparasite may be loosely defined as a fungus existing in intimate association with another from which it derives some or all of its nutrients while conferring to benefit to return. This is an acceptable description of biotrophic mycoparasites, in which there is persistent contact with or occupation of living host cells, but it is inadequate with respect to necro-trophs, by which host cells are killed often before penetration has oc-curred. With a few exceptions, the necrotrophs are the major potential biocontrol agents.

The *sine qua non* of necrotrophic mycoparasitism is contact with another fungus (the target fungus) which leads to death of all or part of the latter. Commonly, penetration of the target fungus occurs at some stage following contact, although death may have already occurred by this time. However, in some instances penetration never takes place even after prolonged con-tact with the dead target species. The target organ (or target phase) may be a hypha or spore or any other kind of vegetative or reproductive structure. Where it is vegetative hypha, abnormalities in growth and branching may occur shortly before contact, these being induced by the action, over relatively small distances, of fungistatic or fungitoxic metabolites ema-nating from the mycoparasite. This kind of antibiosis would seem to be quite distinct from that typical of other competitive interactions in which action is over relatively long distances and results in gross physical and physiological disturbances, and frequently death, well before any contact is possible. However, in natural conditions antibiosis and mycoparasitism may have equally important roles although their precise mode of action is not always clear. Necrotrophic mycoparasites are presumed to acquire endogenous metabolites from their killed target fungi although, at least with respect to parasitised hyphae, evidence for this is almost entirely circumstantial. In addition, any exogenous nutrients formerly accessible to the target fungus become available for sequestration and its territory may be occupied. Captured resources can then be utilised to support further vegetative growth or reproduction.

At any stage during their development, mycoparasites may themselves be subject to attack by parasitic fungi, and this aspect of their biology has tended to be overlooked, despite its obvious relevance to the selection of biological control agents. Indeed, screening procedures have frequently ignored many less obvious aspects of mycoparasitic behaviour, and conclusions regarding biocontrol potential have frequently been reached under controlled and often ecologically unrealistic conditions. However, a thorough knowledge of the various events in the process of mycoparasitism facilitates a more rational choice of test fungi, based on selection methods of greater resolution, and also is of value when devising appropriate delivery systems for them.

9.3 Mycoparasitic interactions

Interactions between mycoparasites and their target fungi occur in four sequential but overlapping phases: target location, recognition, contact and penetration, and nutrient acquisition. The presence, duration and importance of each depend on the fungi involved, whether the mycoparasite is biotrophic or necrotrophic, the target organ being attacked, and the nature of the habitat and its prevailing environmental conditions.

9.3.1 *Hyphal interactions*

Most information on hyphal interaction has been obtained by observing the approach and contact of paired colonies of necrotrophic mycoparasites and target species on nutrient-rich agar. Although behaviour in such conditions *in vitro* may reflect behavior *in vivo*, it should be appreciated that in the majority of situations in which biological control is required to operate, for example in soil or on the rhizoplane or phylloplane, massive opposition of mycelia is an impossibility in the absence of artificial nutrient enrichment. Confrontations would normally be between sparse hyphal systems or, more commonly, to combat between individual hyphae. Examples of typical interactions between the aggressive mycoparasite *Pythium oligandrum* and a number of target species are illustrated in Figs 9.1–9.12.

Some mycoparasites can detect and accurately locate hyphae of target species within their vicinity. For example, hyphae of *Pythium oligandrum* and *Trichoderma hamatum* exhibit directed growth towards those of susceptible fungi, presumably in response to a chemical stimulus produced by the target hypha (Hubbard *et al.*, 1983; Lutchmeah & Cooke, 1984). The proximity of a susceptible hypha to that of a mycoparasite can induce formation of lateral branches on the latter which then show strongly directed growth towards it. Morphogenetic or antibiotic factors released by the mycoparasite may also act on the target hypha to either retard its extension or cause it to produce laterals of limited growth and abnormal morphology.

After contact has been made, the mycoparasite continues to grow over or along the hypha, commonly coiling tightly around it. The degree of

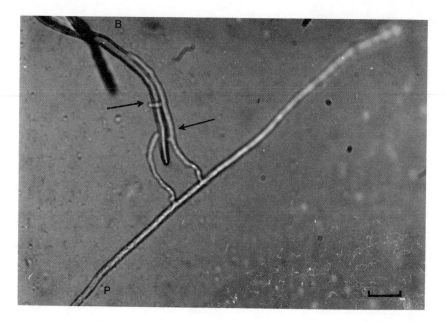

Fig. 9.1. Formation of lateral branches (arrowed) on *Pythium oligandrum* (P) hypha exhibiting directed growth towards an approaching *Botrytis cinerea* hypha (B). Bar = 20 μm.

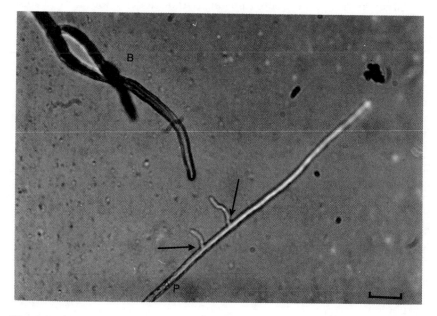

Fig. 9.2. Five minutes later, further lateral branches of *P. oligandrum* (arrowed) effect penetration. Bar = 20 μm.

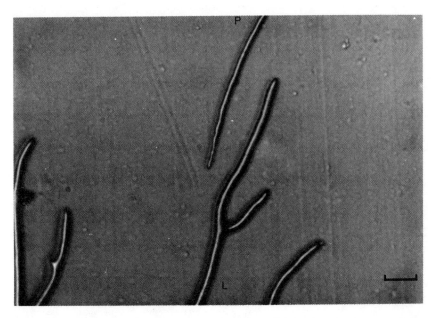

Fig. 9.3. *Pleospora betae* (L) and *P. oligandrum* (P) hyphae in close proximity. Bar = 20 μm.

Fig. 9.4. Same hyphae 10 minutes later. *P. oligandrum* producing laterals (arrowed) which make contact with the target hypha (L). Bar = 20 μm.

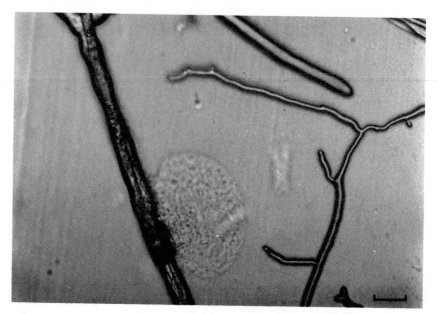

Fig. 9.5. Exit of an intracellular *P. oligandrum* (P) hypha results in lysis of *Sclerotinia sclerotiorum* (S) hypha in which it has grown. Bar = 20 μm.

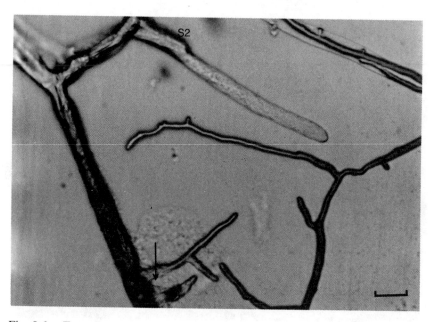

Fig. 9.6. Ten minutes later, initial exit branch continues to grow. *P. oligandrum* exhibits further branching from inside the target hypha (arrowed) and a second *S. sclerotiorum* hypha (S2) loses opacity as a result of penetration. Bar = 20 μm.

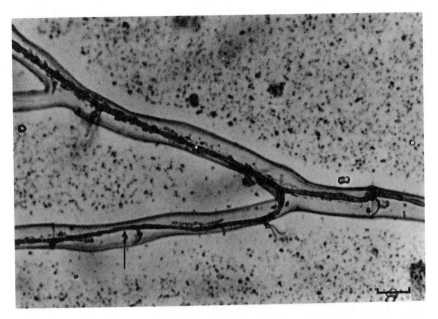

Fig. 9.7. An intracellular *P. oligandrum* hypha (arrowed) growing within a *Botrytis cinerea* hypha. Bar = 100 μm. © GCRI, Littlehampton, UK.

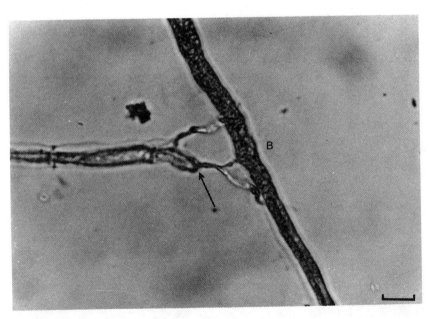

Fig. 9.8. An intracellular *P. oligandrum* hypha exits the tip of a B. *cinerea* hypha (arrowed) and subsequently penetrates a second hypha (B) causing granulation of the cytoplasm. Bar = 100 μm. © GCRI, Littlehampton, UK.

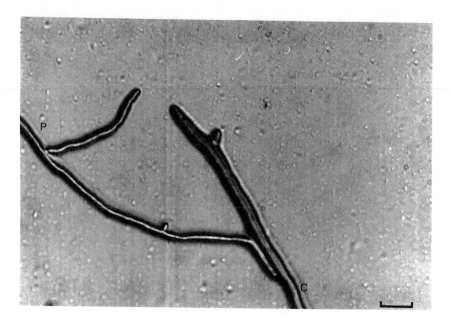

Fig. 9.9. Growing tip of *P. oligandrum* (P) hypha approaches the side of a *Sclerotium cepivorum* (C) hypha. Bar = 20 μm.

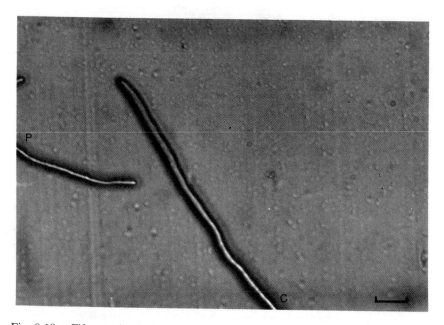

Fig. 9.10. Fifteen minutes later contact occurs between the same two hyphae. Bar = 20 μm.

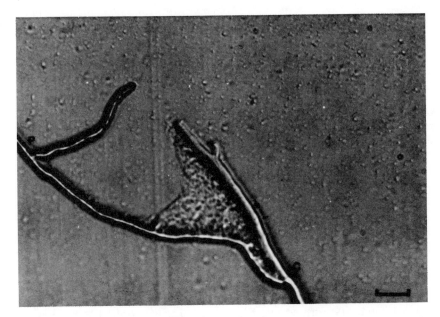

Fig. 9.11. Same hyphae one minute later; penetration of the *S. cepivorum* hypha by a lateral branch formed on the *P. oligandrum* hypha results in lysis of the former. Bar = 20 μm.

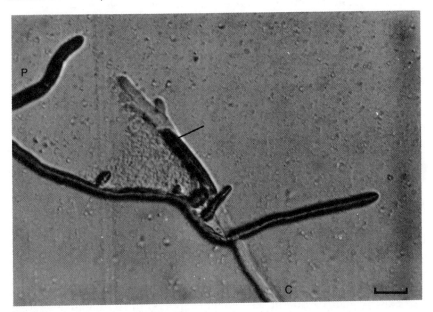

Fig. 9.12. Thirty-five minutes after contact *S. cepivorum*, hypha has lost opacity and ceased to elongate whereas *P. oligandrum* continues to grow forming an intracellular hypha (arrowed). Bar = 20 μm.

coiling depends on the host–parasite combination and can vary widely within a single mycoparasitic species, but there seems to be no correlation between coiling behaviour and mycoparasitic ability (Lifshitz *et al.*, 1984). Although the occurrence of directed growth, contact growth and coiling indicate the possession of well developed location and recognition mechanisms, nothing is yet known of these. However, by analogy with behaviour of plant parasitic fungi it seems likely that location involves the detection of metabolite gradients by the mycoparasite and that post-contact recognition is lectin-mediated (Elad *et al.*, 1983a; Barak *et al.*, 1985).

Cytoplasmic degeneration allowed by death of the target hypha can occur before contact is made, as in *Sclerotium cepivorum* attacked by *Trichoerma harzianum*, or within a few minutes of contact, as with *Pythium oligandrum* and a range of target species (De Oliviera *et al.*, 1984; Lutchmeah & Cooke, 1984). However, in the majority of encounters penetration would seem to be an essential preliminary to, and probable cause of, mortality in parasitised hyphae. This is effected by fine invasive hyphae which may arise either directly from their parent hypha or from an appressorium first formed upon it (Tu, 1980; Elad *et al.*, 1983b; Henis *et al.*, 1984; Kuter, 1984; Lifshitz *et al.*, 1984). Wall-lysing enzymes are brought into play at this stage, and those mycoparasites that have been studied in detail, notably *Pythium nunn* and *Trichoderma* species, possess an array of chitinases, glucanases and proteases which facilitates penetration of both septate and oomycetous fungi. In *Pythium nunn* these enzymes are inducible rather than constitutive, their release being triggered by contact only with susceptible target species (Hadar *et al.*, 1979; Chet & Baker, 1981; Elad *et al.*, 1983b, 1985).

Subsequent to penetration there may be extensive growth within the dead or dying hypha, the contents of which are probably utilised by the mycoparasite. There is, however, little or no direct evidence as to the contribution of any nutrients thus acquired to the development of necrotrophic mycoparasites, either in laboratory culture or in more natural conditions.

9.3.2 *Destruction of propagules*

Many mycoparasites destroy the reproductive or resting phases of plant pathogens and a number are highly adapted in this respect; so that they have little or no ability to attack actively growing vegetative hyphae but depend for survival on successful exploitation of propagules. In above-ground habitats, for instance on leaves and stems, adaptation has generally been towards parasitism of asexual fructifications, although possibilities also exist for destruction of disseminated spores. In soil the major target phases consist of disseminated spores, sexual spores (which are often dormant), sclerotia and, occasionally, resting mycelia. By contrast with the situation for hyphal interactions, studies on this aspect of mycoparasitism *in vitro* have often been supported by appropriate observations *in vivo*.

9.3.2.1 *The phylloplane.* There is a vast and comprehensive literature on competitive interactions between phylloplane fungi but this contains scant evidence for widespread mycoparasitism (Dubos & Bulit, 1981; Blakeman & Fokkema, 1982; Blakeman, 1985). It does, however, occur, most notably within rust sori and on the sporulating mycelium of powdery mildews, and can affect the reproductive potential of these biotrophic plant pathogens, although not perhaps severely.

A small number of necrotrophic mycoparasites associated with one or more spore stages of rust fungi have been described that appear to be ecologically obligately dependent upon them. These include species of the Hyphomycetes *Aphanocladium* and *Tuberculina* and, more familiarly, the Coelomycete *Sphaerellopsis filum* (= *Darluca filum*) which can attack all four spore stages of a wide range of rust genera (Kranz, 1973). Details of infection mechanisms are lacking but what little information there is suggests that rust spores, particularly urediniospores, are entered and destroyed with the aid of wall-degrading enzymes, sporulation of the mycoparasite then taking place on dead host material (Carling *et al.*, 1976; Wicker, 1981). Less specialised mycoparasites, for example *Verticillium lecanii*, *Scytalidium uredinicola* and species of *Alternaria* and *Cladosporium*, can also attack rust pustules via enzymic degradation of spore walls, sometimes accompanied by fungitoxin production, but subsequent penetration of the spores does not always take place (Omar & Heather, 1979; Spencer & Atkey, 1981; Traquair *et al.*, 1984; Grabski & Mendgen, 1985; Srivastava *et al.*, 1985). By contrast with these necrotrophs the Coelomycete *Ampelomyces quisqualis* is apparently biotrophic on powdery mildews although the physiology of the host–mycoparasite relationship is far from clear (Phillip, 1985).

Interactions between mycoparasites and spores of necrotrophic leaf pathogens are even less well documented, detailed information being restricted to a single report on destruction of *Alternaria brassicae* conidia by *Nectria inventa* (Tsuneda & Skoropad, 1978a, b). Mycoparasitic hyphae coil around, form appressoria upon, penetrate and kill conidia in a manner similar to that described above for hyphal interactions.

9.3.2.2 *Soil.* The widest scope for mycoparasitism of propagules occurs within the rhizosphere and in and around decomposing plant residues. Here reproductive and resting structures are generated in great numbers, often to remain *in situ* for a considerable period. In evolutionary terms, the acquisition of the mycoparasitic habit will have been governed by the degree of selective advantage accruing to an ability to exploit any particular kind of propagule. In this regard it does seem to be the case that large, nutrient-rich, persistent organs have afforded the greatest opportunities for the development of mycoparasitism, their destruction commonly being effected by highly specialised fungi which are ecologically dependent on them for survival. This does not, however, preclude mycoparasitism of smaller, less durable structures.

Conidia are the most numerous propagules in soil but the majority quickly lose viability through the process of soil mycolysis. Thus, although *Gliocladium* species infect conidia of some soil-borne plant pathogens *in vitro*, the contribution of mycoparasitism to their disappearance from soil seems likely to be minor (Huang, 1978; Vakili, 1985). Oospores are much less susceptible to mycolysis, and their destruction via mycoparasitism would appear to occur to a correspondingly greater extent; oospores of *Aphanomyces*, *Phytophthora*, *Pythium* and *Sclerospora* species being attacked by a wide range of Chytridiomycetes, Oomycetes and Hyphomycetes (Rao & Pavgi, 1976; Sneh *et al.*, 1977; Hoch & Abawi, 1979; Wynn & Epton, 1979). Dormant mycelia may similarly become targets, and there is, for instance, some evidence for mycoparasitism of resting hyphae of *Rhizoctonia solani* by *Trichoderma lignorum* (Sanford, 1956).

An extensive list has been compiled of species isolated from sclerotia. Although it is possible that many of them participate in the breakdown of sclerotial materials, it is probable that most are secondary colonisers of already damaged tissues and that only a few are strictly mycoparasitic (Jager *et al.*, 1979; Gladders & Coley-Smith, 1980; Howell, 1982; Artigues *et al.*, 1984; Lee & Wu, 1984; McCredie & Sivasthamparam, 1985; Zazzerini & Tosi, 1985). By contrast with the former, the latter are highly destructive of healthy sclerotia and, in at least some situations, may impose a major restraint on disease caused by sclerotium-forming fungi. They may be either facultatively mycoparasitic, as with some *Gliocladium* and *Trichoderma* species, or be obligately so, as with the Coelomycete *Coniothyrium minitans* and the Hyphomycetes *Laterispora brevirama*, *Sporidesmium sclerotivorum* and *Teratosperma oligocladium*, the mycelia of which, *in vivo*, develop only within or immediately around sclerotia.

These fungi penetrate the melanised, toughened, protective rind of the sclerotium and rapidly colonise the medullary tissues. Destruction of sclerotia by *Trichoderma harzianum* and *Coniothyrium minitans* is facilitated by induced glucanases and chitinases (Jones *et al.*, 1974; Elad *et al.*, 1982a,b). The mycoparasite then sporulates upon or within the degraded tissues (Tu, 1980, 1984; Adams & Ayers, 1983; Phillips & Price, 1983). Although all sclerotium mycoparasites so far described have a wide host range, there is some evidence of specificity. The basis for this is as yet undetermined but is possibly related to enzymic competence (Wells *et al.*, 1972; Ayers & Adams, 1981; Elad *et al* 1982a; Artigues *et al.*, 1984).

The specialised habit of *Sporidesmium sclerotivorum* and *Teratosperma oligocladium* is reflected by the fact that *in vivo* their spores only germinate in the presence of sclerotia (Ayers & Adams, 1979; Adams & Ayers, 1983). A further feature of *S. sclerotivorum* is that it does not produce glucanases but stimulates release of these enzymes by sclerotium tissues, which results in their autodigestion. Such control of host metabolism may be effected by haustoria, the presence of which also indicates the bio-trophic nature of this mycoparasite (Bullock *et al.*, 1986).

9.4 Targeting and selection

The general *raison d'être* for attempting to control pests and diseases biologically has been set out many times and needs no repetition here. However, when contemplating the use of mycoparasites a number of theoretical and practical principles must be considered that are especially important for, if not unique to, control of fungi by fungi. Some of these relate to the enormously varied lifestyles both of mycoparasites and of their target species and to the equally assorted habitat conditions under which control is required to operate.

Provided that sufficient details of disease aetiology are known, then a decision can be made as to which developmental phase of the pathogen is to be targeted. This is turn determines the temporal and spatial context of the projected biocontrol system; for example, whether it is to function in the infection court or distant from it, or whether the primary aim is to be prevention of infection or reduction of carry-over of disease once infection has occurred. The effectiveness of any biocontrol measure will then depend on rational screening protocols and the feasibility of delivering the mycoparasite so selected to the environment of the plant pathogen at the correct time, in sufficient quantity, and in an appropriate state of activity.

Two distinct approaches to the initial selection of biocontrol micro-organisms in general have been taken. In the first, isolations have been made from habitats in which the target organism was known to exist but was apparently naturally suppressed; or from which it was absent but where, if introduced, it would probably have thrived. In either case, potential antagonists should be well adapted to the same biotic and abiotic conditions as those normally experienced by the target pathogen (Rouxel *et al.*, 1979; Scher & Baker, 1980; Weller, 1983; Weller & Cook, 1983). The second approach has involved isolating microorganisms from a wide range of habitats irrespective of whether these contained the target pathogen or could possibly have done so (see Cullen & Andrews, 1984; Cook, 1985). Here, there is an obvious risk that, because of their lack of fitness, it may prove difficult subsequently to establish biocontrol agents obtained by this means in habitats that are alien to them. However, isolation procedures apart, and with special reference to mycoparasites, it has not always been fully understood that the fundamental problem is the successful introduction of the biocontrol agent as a new and dynamic component of a stable, pre-existing ecosystem. This holds even where its contribution to ecosystem processes may be transient rather than, as is the ideal, for a relatively extended period. Failure to appreciate this has sometimes resulted in screening of mycoparasites being carried out, both *in vitro* and *in vivo*, under inappropriate conditions. In this regard some general points might be made concerning a number of ecologically relevant biotic and abiotic factors which have not invariably been taken into account.

Where biocontrol is required to be exerted via hyphal interactions, screening has normally involved observations on intermingling mycelia growing on nutrient-rich agars. It has already been pointed out that, in most situations in which biocontrol is sought (except perhaps some nutrient-rich crop residues), conflict will probably be restricted to either highly attenuated mycelia or individual hyphae. This is a consequence of generally low substrate availability in such microbial habitats, nutrient competition for these by the microbial population at large, and the existence within the latter of various kinds of antagonism. In addition, there is good evidence that the opposition of fungi on nutrient-rich media can result in mycoparasitism being expressed by species which are probably not mycoparasitic in nature (Rudakov, 1978). Screening should therefore be carried out under nutrient conditions which at least approximate to those existing in the environment in which biocontrol is to be attempted. Furthermore, when assessing the effect of the mycoparasite on its target species, information should also be sought on the susceptibility of the mycoparasite to antagonisms by major components of the microbial population into which it is to be placed.

As well as employing ecologically unrealistic nutrient levels, screening tests have usually been carried out at a constant temperature, normally in excess of 20°C and under constantly high water potential, both for agar cultures and for some small-scale experiments *in vivo*. Although those are obvious difficulties involved in simulating natural environmental conditions, more attention might be paid to the effects of fluctuating temperature and water potential on fungal development and to possible differences in ecological requirements of mycoparasites and their target species. The fact that, in temperate regions, soil temperatures below surface level remain comparatively low even during summer has implications with respect to regimes used during screening. For example, the sclerotium-destroying species *Coniothyrium minitans* and *Sporidesmium sclerotivorum* develop optimally at 20–25°C with significantly slower growth at 15°C, a temperature which might be considered to be high in relation to average soil temperatures in the sub-surface layers (Turner & Tribe, 1976; Ayers & Adams, 1981b). Similarly most *Trichoderma* species have optima of 25–35°C but develop poorly at low temperatures (Tronsmo & Dennis, 1977; Gladders & Coley-Smith, 1980).

One last problem relating to initial screening is that of strain variability. Usually, very few (commonly only one) isolates of mycoparasite and/or target species are placed in confrontation, so ignoring the possibility of there being a range of interactions between them. The dangers inherent in a too-restricted sampling scheme are well illustrated by *Coniothyrium minitans* in which isolates vary in their ability to destroy sclerotia (Turner & Tribe, 1976). This may also be in part due to the properties of the sclerotia under test in that their survival ability varies widely depending on whether they have been produced on agar or on living or dead plant tissues, or have been subjected to desiccation stress (Smith, 1972; Trutmann *et al.*, 1980; Coley-Smith, 1985).

The culminating phase of the targeting and selection process is the development and evaluating of suitable delivery systems for the mycoparasite. The nature of available delivery systems and their effectiveness in controlling disease are dealt with below, but it is pertinent here to outline some principles which will determine their eventual successful use. It must be emphasised that judgement of the actual or potential success of a system should not be based solely on the degree of disease control or pathogen population reduction so obtained: this is to take too sanguine a view of the efficacy of biological methods. It is equally important to monitor the fate of the biocontrol agent and to assess the feasibility of practical use of the delivery system. Aspects of post-application behaviour studied should embrace not only the persistence of the mycoparasite but also, for instance where it is introduced into soil, its possible effects on other beneficial fungi (for example mycorrhizal species). Other feasibility factors to be considered include the shelf-life of the delivery product, its capability to control disease when applied in agriculturally practicable amounts, and its potential for possible integration with existent disease management programmes. Finally, the crux will be the costing of the biological system *vis-à-vis* other available methods and against a background of continually evolving chemical control measures and increasing use of genetic engineering and cloning to accelerate production of resistant cultivars.

9.5 Microbiomes, delivery systems and disease control

Biocontrol using mycoparasites has been attempted in widely different situations using a corresponding variety of methods. Appraisal of the literature leads to one firm conclusion: that demonstration of control is not enough for practical purposes unless there are supporting data of the kind outlined above. Unfortunately these are often lacking, which makes for difficulty in predicting the likely success of any particular method on a realistic scale. While seeking to preserve an open view, attention here will be focused on selected examples which, within the limitations of current assessment, seem to have clear potential for future practical application.

A convenient ecological framework in which to examine biocontrol systems is that of the microbiome. This may be defined as a characteristic microbial community occupying a reasonably well defined habitat which has distinct physico-chemical properties. The term thus not only refers to the microorganisms involved but also encompasses their theatres of activity. In relation to fungal diseases of crops and their control, major microbiomes are the phylloplane, spermosphere, rhizosphere and rhizoplane, and numerous kinds of plant residues persisting on or in the soil. Mention should also be made of the wood of standing or felled trees as microbiomes where biocontrol of forest diseases using fungi has been achieved. However, in most cases competitive interactions other than mycoparasitism seem to be of greater importance.

9.5.1 *Aerial microbiomes*

General prospects for biocontrol of foliage diseases have been outlined recently and it is quite clear that, in the majority of instances, the current state is still one of exploration rather than exploitation (Blakeman & Fokkema, 1982; Windels & Lindow, 1985). Major constraints on the development of dependable methods are the rapidity with which phylloplane conditions can fluctuate in the field plus the inability of many antagonists to restrain pathogenesis once penetration of the leaf has taken place. In addition, it does not seem to be widely enough appreciated that the success of leaf pathogens reflects the fact that they are highly adapted to abiotic phylloplane conditions and also are not affected catastrophically, in normal circumstances, by most of the antagonists they are likely to encounter. It follows from this that any potential antagonists drawn from other microbiomes for introduction to the phylloplane will probably be poorly adapted unless either a degree of fitness has first been obtained via genetic manipulation or, as also can be envisaged for resident antagonists, the environment can be modified to their advantage. This applies especially to mycoparasites, where those difficulties encountered in exploiting even apparently well adapted species are exemplified by the situation with regard to rust diseases.

It has been argued, on theoretical grounds using mathematical models, that at low cereal rust densities the action of natural enemies can delay epidemic development and so substantially reduce yield losses. On this basis it has further been proposed that artifical boosting of the natural enemy complex, with particular respect to *Sphaerellopsis filum*, may prove to be effective (Sidorova & Khasanov, 1975; Hau & Kranz, 1978; Fleming, 1980). Such suggestions are founded on a total misunderstanding of the biology of this specialised fungus, which can probably establish itself as a persistent leaf inhabitant only after development of rust pustules (Swendsrud & Calpouzos, 1972). Although it may indeed subsequently reduce disease severity and dissemination by destroying rust spores, its initial ability to invade pustules may be severely affected by changes in phyllo-plane conditions, particularly decreases in water potential (Ramakrishman & Narasimhalu, 1941; Prasada, 1948). A further salient characteristic of *Sphaerellopsis filum* is that its main means of reproduction is through pycnidial conidia released in a mucilaginous matrix. These wet spore masses are then dispersed mainly via rain splash, so that rates of spread from both natural and artificially induced infection foci are likely to be inadequate to match those of rust diseases in either cereals or other crops (Kuhlman *et al.*, 1978).

Other, less specialised, species are likely to prove more useful in rust control. For instance, in field conditions, infection by *Scytalidium uredini-cola* of loblolly pine (*Pinus taeda*) stem galls caused by *Cronartium quercinum* f.sp. *fusiforme* reduced aeciospore production by 72% and aeciospore viability by 28% (Kuhlmann, 1981). However, data on rates of spread, survival, and conditions for infection by this mycoparasite in nature are lacking.

The most fruitful province for rust control, and possibly for control of foliage diseases of most kinds, is within protected crops. Here, there is the possibility of regulating environmental factors in glasshouses and other production units so as to favour mycoparasitic activity. One antagonist which appears to have exceptional promise is *Verticillium lecanii*, a common glasshouse inhabitant that has been used successfully to control aphid and whitefly infestations. Its conidia can be incorporated into commercial spray formulations (Hall, 1982; Kanangaratnam *et al.*, 1982). This fungus is remarkable, and perhaps unique, in that it is not only facultatively entomo-genous but is also mycoparasitic, being highly destructive of urediniospores of the carnation rust, *Uromyces dianthi* (Spencer, 1980; Spencer & Atkey, 1981). It can act at the penetration stage of the rust so reducing infection and, in addition, if applied after uredinia have begun to appear, it inter-rupts the development of further pustules. The latter effect may be due to either reduced viability of urediniospores in mycoparasitised uredinia or to antibiosis.

Beyond protected environments, vagaries in aerial climate create problems for targeting pathogens at the crucial pre-penetration and pene-tration phases of their development. It is probably infeasible to establish most mycoparasites in the infection court prior to the arrival of their target pathogens. Rather, they must be delivered at approximately the same time as pathogenic propagules are being deposited there. Such a degree of precision requires accurate disease forecasting and, if the effects of mycoparasites are non-persistent–as indeed seems to be the case, regular applications throughout the risk period. Even then, the rapidity with which pathogenic spores effect penetration may allow escape into host tissues, from which the mycoparasite may be excluded, so that disease still develops.

This kind of behaviour has been observed during studies on biocontrol of *Alternaria brassicae* (black spot) on oilseed rape by *Nectria inventa* (Tsundea & Skoropad, 1978a, b). On attached rape leaves, conidia of *Alternaria brassicae* germinate more quickly than those of *Nectria inventa* and the pathogen so avoids mycoparasitism. However, black spot disease results in early defoliation, and on detached leaves *Nectria inventa* suppresses both vegetative growth and sporulation of the pathogen. Carryover of infection could thus greatly be reduced or prevented although this possibility has not been tested in the field. Since *Nectria inventa* is a vigorous saprotroph which is probably well able to maintain itself on decaying litter in field conditions, such an approach to black spot control holds promise.

In wider terms, and for the reasons outlined above, disease control by prevention of inoculum carry-over might prove to be more practicable than direct intervention in the infection court, and certainly requires further investigation. A recently described mycoparasite, *Cladobotryum amazonense*, is of great interest here. It attacks the basidiocarps of the agaric *Crinipellis perniciosa*, a major pathogen of cocoa, preventing normal hymenial development and reducing basidiospore release; direct mycoparasitism being complemented by release of fungitoxic metabolites which plasmolyse basidiospores (Bastos *et al.*, 1981).

9.5. Soil microbiomes.

It is fairly certain that the best prospects for exploitation of mycoparasites lie in control of soil-borne diseases, and this is reflected in the variety of approaches currently being made to this problem. Despite the notorious biological buffering capacity of soils, a number of techniques are now available which permit the successful introduction of mycoparasite populations. By contrast with the aerial environment, soil conditions are relatively stable and fluctuations in them are relatively slow. In addition it is often possible to regulate some important soil factors, for example pH, water potential and organic matter content, so as to favour the operation of biocontrol systems.

Two basic strategies have been adopted for delivering mycoparasites to soil microbiomes. The first involves localising biocontrol inoculum around the microbiome which is to be protected. This may be achieved by techniques such as pelleting or seed coating or treating other materials used in plant propagation. Here, it is visualised that concurrent development of plant and mycoparasite will result in the latter coming to occupy the spermosphere, rhizoplane and rhizosphere, although perhaps only during the early stages of plant growth. The second strategy relies on distributing biocontrol inoculum evenly throughout the bulk soil volume. This facilitates continuous protection close to the plant as it develops, and may also being about destruction of pathogenic propagules at some distance from it.

Numerous procedures have been used to incorporate mycoparasites into seed pelleting or seed coating materials. On a glasshouse scale *Pythium*-induced damping-off and seed rots in pea have been successfully controlled using *Trichoderma harzianum* and *T. koningii* in adhesive seed coatings of polyvinyl alcohol, or in fluid drilling gels such as hydroxymethyl cellulose (Polysurf C) or glucan gum (Polytran N). Direct evidence has been obtained for mycoparasitism of *Pythium ultimum* on the pea seed surface (Hubbard *et al.*, 1983; Hadar *et al.*, 1984). Similarly the aggressive mycoparasite *Pythium oligandrum* has been pelleted, as oospores, with seed in clay carrier to control *Pythium*-induced damping-off in sugar beet and *Mycocentrospora acerina* infection of seedling carrots (Lutchmeah & Cooke, 1985). *Pythium oligandrum* oospores can survive for more than 12 months in stored pelleting material and for up to 18 months in soil (Al-Hamdani *et al.*, 1983). The exciting possibility of integration of biological and chemical control methods is indicated by the observation that seed pelleting with ultraviolet induced mutants of *Trichoderma* species, which were fungicide-resistant, achieved control of *Pythium*-incited damping-off in pea (Papavizas *et al.*, 1982; Papavizas & Lewis, 1983).

Whilst such treatments can control seed and seedling diseases, maintenance of protection beyond the seedling stage requires establishment of the mycoparasite around the hypocotyl and within the ever-expanding rhizosphere. Some mycoparasites, notably species of *Gliocladium* and *Trichoderma*, are deficient in these respects in that they fail to travel down roots (Papavizas, 1981; Chao *et al.*, 1986). On the other hand *Verticillium biguttatum* is able to spread from coated seed tubers of potato to sprouts and stolons, protecting them from infection by *Rhizoctonia solani* and producing a 60% reduction in sclerotium production by the pathogen (Velvis & Jager, 1983; Jager & Velvis, 1984, 1985).

Effects of distributing mycoparasites throughout soil have been studied chiefly for *Gliocladium* and *Trichoderma* species. Early methods of application achieved control, but the huge amounts of inoculum required to be applied were agriculturally unrealistic (Wells *et al.*, 1972; Backman & Rodriguez-Kabana, 1975). More recently a number of practicable systems have been developed, using dried fermenter biomass accommodated within various carriers, or by bulking out the products of solid fermentation with inert materials. A useful development has been the monitoring of the mycoparasite population after its addition to soil. In a major study large quantities of *Gliocladium* and *Trichoderma* biomass were obtained from fermentation of molasses, and the dried product, consisting of hyphae, conidia and chlamydospores, survived well at $-5°C$ and $5°C$ when mixed with pyrophyllite carrier. After addition to soil the mycoparasites reproduced, as indicated by a rise in the number of colony forming units (CFU) present per unit weight of soil, suggesting their firm establishment there (Papavizas *et al.*, 1984). A different method has used bran–sand mixtures inoculated with conidia which, after a 3-day incubation period, were mixed with soil (Lewis & Papavizas, 1984). These, largely mycelial, applications gave rise to a $\times 10^6$ increase in CFU in several different soils within 3 weeks. Survival of the target pathogen *Rhizoctonia solani* in soil was at the same time reduced by at least 50%; such inoculum also reduced damping-off in cotton, suger beet and radish in artificially infested soil. Mycoparasitism of *Rhizoctonia solani* by *Trichoderma* species was directly observed in soil (Lewis & Papavizas, 1985a). An important recent advance has been the incorporation of fermenter biomass of *Gliocladium* and *Trichoderma* into alginate pellets which can then be added to soil after being bulked with either bran or kaolin (Fravel *et al.*, 1985; Lewis & Papavizas, 1985b).

Increases in CFU following addition of mycoparasites of this kind to soil are probably the result of sporulation rather than a reflection of vegetative development. It can be anticipated that, in the absence of repeated additions of fresh inoculum or of energy-rich substrata to promote growth and further sporulation, the mycoparasite population will decline, possibly rapidly, in the face of biotic and abiotic stresses. Among the former, microbial antagonisms will probably play a major part in reducing the number and effectiveness of vegetative hyphae, conidia and resting spores. Reducing the population of soil microorganisms by chemical treatments prior to, or coincident with, introduction of the mycoparasite therefore might, in some circumstances, improve subsequent biocontrol. For example, soil fumigation using methyl bromide at 200 kg/ha failed to control *Rhizoctonia solani*, but, combined with an application of *Trichoderma harzianum*, control was achieved equivalent to that obtained using the recommended methyl bromide dosage of 500 kg/ha (Strashnow *et al.*, 1985). The resistance of *Trichoderma harzianum* to many fungicides has enabled non-sterile straw to be used for production of biocontrol inoculum. This species will grow on straw treated with vinclozolin, and addition of such material to soil has resulted in control of *Sclerotinia minor* on lettuce in field conditions (Davet *et al.*, 1981; Davet & Martin, 1985).

A little-explored aspect of antagonism, which may involve mycoparasitism, is the suppression of pathogens in potting composts and horticultural container media. Peat-based container media amended with composted hardwood bark developed dense populations of *Trichoderma* species associated with reduced damping-off caused by *Rhizoctonia solani* (Kuter *et al.*, 1983). Similarly, addition of *Trichoderma viride* conidia to steamed plant growth media produced high populations of the biocontrol agent and suppressed artificial infestations of the wilt pathogen *Fusarium oxysporum* f.sp. *chrysanthemi* (Marois & Locke, 1985).

One final area in which to find encouragement concerns the mycoparasitic destruction of sclerotia. These are relatively large structures. often massive, and usually possess great longevity. They constitute microbiomes favourable to a range of microorganisms by virtue of their size, degree of permanence, and the fact that they leak nutrients, especially when subjected to wetting/drying cycles. Mention has already been made of specific sclerotium-destroying mycoparasites, and there is good evidence that some of these can be introduced into the field with some success. For example, biomass of *Coniothyrium minitans* grown on a barley–rye–sunflower seed mixture and applied to sunflower seed rows reduced *Sclerotinia* wilt by 30% over a two-year period (Huang, 1980). This indicates that *Coniothyrium* can exhibit a degree of self-maintenance in field conditions. Of the mycoparasites so far examined *Sporodesmium sclerotivorum* shows perhaps the greatest potential. A single application of this species, grown on nonsterile sand containing 1% w/v live sclerotia of *Sclerotinia minor*, to plots at rates of 0.2 and 2.2 kg/m ($= 10^2$ and 10^3 macro-conidia per gram of soil) reduced lettuce drop by 40–80% in four successive crops over a two-year period (Adams & Ayers, 1982; Adams *et al.*, 1984). Spores of the mycoparasite germinated in the presence of sclerotia and, after colonisation, its hyphae grew from sclerotium to sclerotium through the soil (Ayers & Adams 1979, 1981). It has a well-developed ability to infect sclerotia in natural conditions, utilising them as a protected nutrient source, and subsequently producing large numbers of persistent macro-conidia. It has been grown on a commercial scale using a vermiculite-based medium, and the results of wider trials are awaited (Adams *et al.*, 1984).

9.6 Conclusions

Given appropriate conditions and suitable application methods mycoparasites can control many diseases. But the question remains as to how soon and on what scale they might be expected to come into commercial use in agriculture and horticulture. Two obstacles must be cleared before any such predictions can be made with confidence. First, more information is required on the behaviour of mycoparasites in the field. In particular, knowledge of the short- and long-term population dynamics of mycoparasite and target pathogen is essential to any proposed control programme. Short-term success may not be matched by longer-term benefits; interfering in stable microbial ecosystems may not be as simple a procedure as might be assumed. For instance, initial control of *Rhizoctonia solani* populations in soil by addition of *Laetisaria arvalis* is good; but rapid recovery follows as the natural antagonist–target equilibrium is regained (Allen *et al.*, 1985). Secondly, and irrespective of current concern with pesticides and the environment, it must be clearly demonstrated that biological control not only works but that it can be more cost-effective than available chemical control. Only then will it be possible to draw on the financial resources of the agro-chemical industry which are necessary to bring theory into practice.

References

Adams, P.B. & Ayers, W.A. (1982). Biological control of sclerotinia lettuce drop in the field by *Sporidesmium sclerotivorum*. *Phytopathology*, **72**, 485–8.

Adams, P.B. & Ayers, W.A. (1983). Histological and physiological aspects of infection of sclerotia of *Sclerotinia* species by two mycoparasites. *Phytopathology*, **73**, 1072–6.

Adams, P.B., Marois, J.J. & Ayers, W.A. (1984). Population dynamics of the mycoparasite *Sporidesmium sclerotivorum*, and its host, *Sclerotinia minor*, in soil. *Soil Biology and Biochemistry*, **16**, 627–33.

Al-Hamdani, A.M., Lutchmeah, R.S. & Cooke, R.C. (1983). Biological control of *Pythium ultimum*-induced damping-off by treating cress seed with the mycoparasite *Pythium oligandrum*. *Plant Pathology*, **32**, 449–54.

Allen, M.F., Boosalis, M.G., Kerr, E.D., Muldoon, A.E. & Larsen, H.J. (1985). Population dynamics of sugar beets, *Rhizoctonia solani*, and *Laetisaria arvalis*: responses of a host, plant pathogen, and hyperparasite to perturbation in the field. *Applied and Environmental Microbiology*, **50**, 1123–7.

Artigues, M., Davet, P. & Roure, C. (1984). Comparison des aptitudes parasitaires de clones de *Trichoderma* vis-à-vis de quelques champignons a sclerotes. *Soil Biology and Biochemistry*, **16**, 413–17.

Ayers, W.A. & Adams, P.B. (1979). Mycoparasitism of sclerotia of *Sclerotinia* and *Sclerotium* species by *Sporidesmium sclerotivorum*. *Canadian Journal of Microbiology*, **25**, 17–23.

Ayers, W.A. & Adams, P.B. (1981). Mycoparasitism and its application to biological control of plant diseases. In: *Biological Control in Crop Protection*, ed. G.C. Papavizas, pp. 91–103. Allanheld & Osmun, Totowa, NJ.

Backman, P.A. & Rodriguez-Kabana, R. (1975). A system for the growth and delivery of biological control agents to the soil. *Phytopathology*, **65**, 819–21.

Barak, R., Elad, Y., Mirelman, D. & Chet, I. (1985). Lectins: A possible basis for specific recognition in the interaction of *Trichoderma* and *Sclerotium rolfsii*. *Phytopathology*, **75**, 458–62.

Bastos, C.N., Evans, H.C. & Samson, R.A. (1981). A new hyperparasitic fungus, *Cladobotryum amazonense*, with potential for control of fungal pathogens of cocoa. *Transactions of the British Mycological Society*, **77**, 273–8.

Blakeman, J.P. (1985). Ecological succession of leaf surface microorganisms in relation to biological control. In: *Biological Control on the Phylloplane*, eds C.E. Windels & S.E. Lindow, pp. 6–30. American Phytopathological Society, St. Paul, Minn.

Blakeman, J.P. & Fokkema, N.J. (1982). Potential for biological control of plant diseases on the phylloplane. *Annual Review of Phytopathology*, **20**, 167–92.

Bullock, S., Adams, P.B., Willetts, M.J. & Ayers, W.A. (1986). Production of haustoria by *Sporidesmium sclerotivorum* in sclerotia of *Sclerotinia minor*. *Phytopathology*, **76**, 101–3.

Carling, D.E., Brown, M.F. & Millikan, D.F. (1976). Ultrastructural examination of the *Puccinia graminis–Darluca filum* host–parasite relationship. *Phytopathology*, **66**, 419–22.

Chao, W.L., Nelson, E.B., Harman, G.E. & Hoch, H.C. (1986). Colonization of the rhizosphere by biological control agents applied to seeds. *Phytopathology*, **76**, 60–5.

Chet, I. & Baker, R. (1981). Isolation and biocontrol potential of *Trichoderma hamatum* from soil naturally suppressive to *Rhizoctonia solani*. *Phytopathology*, **71**, 286–90.

Coley-Smith, J.R. (1985). Methods for the production and use of sclerotia of *Sclerotium cepivorum* in field germination studies. *Plant Pathology*, **34**, 380–4.

Cook, R.J. (1985). Biological control of plant pathogens: theory to application. *Phytopathology*, **75**, 25–9.

Cullen, D. & Andrews, J.H. (1984). Epiphytic microbes as biological control agents. In: *Plant–Microbe Interactions. Molecular and Genetic Perspectives*, Vol. 1, eds T. Kosuge & E.W. Nester, pp. 381–99. Macmillan, London.

Davet, P., Artigues, M. & Martin, C. (1981). Production en conditions non aseptiques d'inoculum de *Trichoderma harzianum* Rifai pour des essais de lutte biologique. *Agronomie*, **1**, 933–6.

Davet, P. & Martin, C. (1985). Effets de traitements fongicides aux imides cycliques sur les populations de *Sclerotinia minor* dans le sol. *Phytopathologische Zeitschrift*, **112**, 7–16.

De Oliveira, V.L., Bellei, M. & Borges, A.C. (1984). Control of white rot of garlic by antagonistic fungi under controlled environmental conditions. *Canadian Journal of Microbiology*, **30**, 884–9.

Dubos, B. & Bulit, J. (1981). Filamentous fungi as biocontrol agents on aerial plant surfaces. In: *Microbial Ecology of the Phylloplane*, ed. J.P. Blakeman, pp. 353–81. Academic Press, London.

Elad, Y., Barak, R. & Chet, I. (1983a). Possible role of lectins in mycoparasitism. *Journal of Bacteriology*, **154**, 1431–5.

Elad, Y., Barak, R., Chet, I. & Henis, Y. (1983b). Ultrastructural studies of the interaction between *Trichoderma* spp. as plant pathogenic fungi. *Phytopathologische Zeitschrift*, **107**, 168–75.

Elad, Y., Chet, I. & Henis, Y. (1982a). Degradation of plant pathogenic fungi by *Trichoderma harzianum*. *Canadian Journal of Microbiology*, **28**, 719–25.

Elad, Y., Hadar, Y. & Henis, Y. (1982b). Prevention with *Trichoderma harzianum* Rifai aggr., of reinfestation by *Sclerotium rolfsii* Sacc. and *Rhizoctonia solani* Kühn of soil fumigated with methyl bromide, and improvement of disease control in tomatoes and peanuts. *Crop Protection*, **11**, 199–211.

Elad, Y., Lifshitz, R. & Baker, R. (1985). Enzymatic activity of the mycoparasite *Pythium nunn* during interaction with host and non-host fungus. *Physiological Plant Pathology*, **27**, 131–48.

Fleming, R.A. (1980). The potential for control of cereal rust by natural enemies. *Theoretical Population Biology*, **18**, 374–95.

Fravel, D.R., Marois, J.J., Lumsen, R.D. & Connick, W.J. (1985). Encapsulation of potential biocontrol agents in an alginate–clay matrix. *Phytopathology*, **75**, 774–7.

Gladders, P. & Coley-Smith, J.R. (1980). Interactions between *Rhizoctonia tuli-parum* sclerotia and soil microorganisms. *Transactions of the British Mycological Society*, **74**, 579–86.

Grabski, G.C. & Mendgen, K. (1985). Einsatz von *V. lecanii* als biologisches Schädlingsbekämpfungsmittel gegen den Bohnenrostpilz *V. appendiculatus* var. *appendiculatus* im Feld und im Gewächshaus. *Phytopathologische Zeitschrift*, **113**, 243–51.

Hadar, Y., Chet, I. & Henis, Y. (1979). Biological control of *Rhizoctonia solani* damping-off with wheat bran culture of *Trichoderma harzianum*. *Phytopathology*, **69**, 64–8.

Hadar, Y., Harman, G.E. & Taylor, A.G. (1984). Evaluation of *Trichoderma koningii* and *T. harzianum* from New York soils for biological control of seed rot caused by *Pythium* spp. *Phytopathology*, **74**, 106–10.

Hall, R.A. (1982). Control of whitefly *Trialeurodes vaporariorum* and cotton aphid, *Aphis gossypii*, in glasshouses by two isolates of the fungus, *Verticillium lecanii*. *Annals of Applied Biology*, **101**, 1–11.

Hau, B. & Kranz, J. (1978). Model computations on the effectiveness of the hyperparasite *Eudarluca caricis* against rust epidemics. *Zeitschrift für Pflan-zenkrankheiten und Pflanzenschutz*, **85**, 137–41.

Henis, Y., Lewis, J.A. & Papavizas, G.C. (1984). Interactions between *Sclerotium rolfsii* and *Trichoderma* spp.: relationship between antagonism and disease con-trol. *Soil Biology and Biochemistry*, **16**, 391–5.

Hoch, H.C. & Abawi, G.S (1979). Mycoparasitism of oospores of *Pythium ultimum* by *Fusarium merismoides*. *Mycologia*, **71**, 621–5.

Howell, C.R. (1982). Effect of *Gliocladium virens* on *Pythium ultimum*, *Rhi-zoctonia solani*, and damping-off of cotton seedlings. *Phytopathology*, **72**, 496–8.

Huang, H.C. (1978). *Gliocladium catenulatum*: hyperparasite of *Sclerotinia sclero-tiorum* and *Fusarium* species. *Canadian Journal of Botany*, **56**, 2243–6.

Huang, H.C. (1980). Control of *Sclerotinia* wilt of sunflower by hyperparasites. *Canadian Journal of Plant Pathology*, **2**, 26–32.

Hubbard, J.P., Harman, G.E. & Hadar, Y. (1983). Effect of soilborne *Pseudo-monas* spp. on the biological control agent, *Trichoderma hamatum*, on pea seeds. *Phytopathology*, **73**, 655–9.

Jager, G., Ten Hoopen, A. & Velvis, H. (1979). Hyperparasites of *Rhizoctonia solani* in Dutch potato fields. *Netherlands Journal of Plant Pathology*, **85**, 253–68.

Jager, G. & Velvis, H. (1984). Biological control of *Rhizoctonia solani* on potatoes by antagonists. 2. Sprout protection against soil-borne *R. solani* through soil inoculation with *Verticillium biguttatum*. *Netherlands Journal of Plant Pathology*, **90**, 29–33.

Jager, G. & Velvis, H. (1985). Biological control of *Rhizoctonia solani* on potatoes by antagonists. 4. Inoculation of seed tubers with *Verticillium biguttatum* and other antagonists in field experiments. *Netherlands Journal of Plant Pathology*, **91**, 49–63.

Jones, D., Gordon, A.H. & Bacon, J.S.D. (1974). Cooperative action by endo- and exo-B-(1–3)-glucanases from parasitic fungi in the degradation of cell wall glucans of *Sclerotinia sclerotiorum*. *Biochemical Journal*, **140**, 47–55.

Kanangaratnam, P., Hall, R.A. & Burges, H.D. (1982). Control of glasshouse whitefly, *Trialeuroides vaporariorum* by an 'aphid' strain of the fungus *Verti-cillium lecanii*. *Annals of Applied Biology*, **100**, 213–19.

Kranz, J. (1973). A host list of the rust parasite *Endarluca caracis* (Fr.) O. Erickss. *Nova Hedwigia*, **24**, 169–80.

Kuhlman, E.G. (1981). Mycoparasitic effects of *Scytalidium uredinicola* on aeciospore production and germination of *Cronartium quercinum* f. sp. *fusiforme*. *Phytopathology*, **71**, 186–8.

Kuhlman, E.G., Matthews, F.R. & Tillerson, H.P. (1978). Efficacy of *Darluca filum* for biological control of *Cronartium fusiforme* and *C. strobilinum*. *Phytopathology*, **68**, 507–11.

Kuter, G.A. (1984). Hyphal interactions between *Rhizoctonia solani* and some *Verticillium* species. *Mycologia*, **76**, 936–48.

Kuter, G.A., Nelson, E.B., Hoitink, H.A.J. & Madden, L.V. (1983). Fungal populations in container media amended with composted hardwood bark suppressive and conducive to *Rhizoctonia* damping-off. *Phytopathology*, **73**, 1450–6.

Lee, Y.A. & Wu, W.S. (1984). The antagonism of *Trichoderma* spp. and *Gliocladium virnes* against *Sclerotinia sclerotiorum*. *Plant Protection Bulletin (Taiwan ROC)*, **26**, 293–304.

Lewis, J.A. & Papavizas, G.C. (1984). A new approach to stimulate population proliferation of *Trichoderma* species and other potential biocontrol fungi introduced into natural soils. *Phytopathology*, **74**, 1240–4.

Lewis, J.A. & Papavizas, G.C. (1985a). Effect of mycelial preparations of *Trichoderma* and *Gliocladium* on populations of *Rhizoctonia solani* and the incidence of damping-off. *Phytopathology*, **75**, 812–17.

Lewis, J.A. & Papavizas, G.C. (1985b). Characteristics of alginate pellets formulated with *Trichoderma* and *Gliocladium* and their effect on the proliferation of the fungi in soil. *Plant Pathology*, **34**, 571–7.

Lifshitz, R., Dupler, M., Elad, Y. & Baker, R. (1984). Hyphal interactions between a mycoparasite *Pythium nunn* and several soil fungi. *Canadian Journal of Microbiology*, **30**, 1482–7.

Lutchmeah, R.S. & Cooke, R.C. (1984). Aspects of antagonism by the mycoparasite *Pythium oligandrum*. *Transactions of the British Mycological Society*, **83**, 696–700.

Lutchmeah, R.S. & Cooke, R.C. (1985). Pelleting of seed with the antagonist *Pythium oligandrum* for biological control of damping-off. *Plant Pathology*, **34**, 528–31.

Marois, J.J. & Locke, J.C. (1985). Population dynamics of *Trichoderma viride* in steamed plant growth medium. *Phytopathology*, **75**, 115–18.

McCredie, T.A. & Sivasthamparam, K. (1985). Fungi mycoparasitic on sclerotia of *Sclerotinia sclerotiorum* in some Western Australian soils. *Transactions of the British Mycological Society*, **84**, 736–9.

Omar, M. & Heather, W.A. (1979). Effect of saprophytic phylloplane fungi on germination and development of *Melampsora larici-populina*. *Transactions of the British Mycological Society*, **72**, 225–31.

Papavizas, G.C. (1981). Survival of *Trichoderma harzianum* in soil and in pea and bean rhizospheres. *Phytopathology*, **71**, 121–5.

Papavizas, G.C., Dunn, M.T., Lewis, J.A. & Beagle-Ristaino, J. (1984). Liquid fermentation technology for experimental production of biocontrol fungi. *Phytopathology*, **74**, 1171–5.

Papavizas, G.C. & Lewis, J.A. (1983). Physiological and biocontrol characteristics of stable mutants of *Trichoderma viride* resistant to MBC fungicides. *Phytopathology*, **75**, 407–11.

Papavizas, G.C., Lewis, J.A. & Abd-El Moity, T.H. (1982). Evaluation of new biotypes of *Trichoderma harzianum* for tolerance to benomyl and enhanced biocontrol properties. *Phytopathology*, **72**, 126–32.

Phillip, D.W. (1985). Extracellular enzymes and nutritional physiology of *Ampelomyces quisqualis* CES., hyperparasite of powdery mildew, *in vitro*. *Phytopathologische Zeitschrift*, **107**, 193–203.

Phillips, A.J.L. & Price, K. (1983). Structural aspects of the parasitism of sclerotia of *Sclerotinia sclerotiorum* (Lib.) de Bary by *Coniothyrium minitans*. *Phytopathologische Zeitschrift*, **107**, 193–203.

Prasada, R. (1948). *Darluca filum* (Biv.) Cast, hyperparasite of *Puccinia graminis* and *Puccinia triticina* in the greenhouse. *Current Science*, **17**, 215–16.

Ramakrishman, T.S. & Narasimhalu, I.L. (1941). The occurrence of *Darluca filum* (Biv.) Cast. on cereal rusts in South India. *Current Science*, **10**, 290–1.

Rao, N.N.R. & Pavgi, M.S. (1976). Sclerotial mycoflora and its role in natural biological control of 'white-rot' disease. *Plant and Soil*, **43**, 509–13.

Rouxel, F., Alabouvette, C. & Louvet, J. (1979). Recherches sur la resistance des sols aux maladies. IV. Mise en evidence du rôle des *Fusarium* autochtones dans la resistance d'un sol a la *Fusariose* vasculaire du melon. *Annales des Phytopathologie*, **11**, 199–207.

Rudakov, O.L. (1978). Physiological groups of mycophilic fungi. *Mycologia*, **70**, 130–59.

Sanford, G.B. (1956). Factors influencing formation of sclerotia by *Rhizoctonia solani*. *Phytopathology*, **46**, 281–4.

Scher, F.M. & Baker, R. (1980). Mechanism of biological control in a *Fusarium* suppressive soil. *Phytopathology*, **70**, 412–17.

Sidorova, I.I. & Khasanov, B.A. (1975). Artificial infection of the causative agent of stem rust of wheat with the hyperparasite *Darluca filum*. *Moscow University Biological Sciences Bulletin*, **30**, 136–8.

Smith, A.M. (1972). Biological control of fungal sclerotia in the soil. *Soil Biology and Biochemistry* **4**, 131–4.

Sneh, B., Humble, S.J. & Lockwood, J.L. (1977). Parasitism of oospores of *Phytophthora megasperma* var. *sojae*, *P. cactorum*, *Pythium* sp. and *Aphanomyces euteiches* by oomycetes, chytridiomycetes, hyphomycetes, actinomycetes and bacteria. *Phytopathology*, **67**, 622–8.

Spencer, D.M. (1980). Parasitism of carnation rust (*Uromyces dianthi*) by *Verticillium lecanii*. *Transactions of the British Mycological Society*, **74**, 191–4.

Spencer, D.M. & Atkey, P.T. (1981). Parasitic effects of *Verticillium lecanii* on two rust fungi. *Transactions of the British Mycological Society*, **77**, 535–42.

Srivastava, A.K., Defago, G. & Kern, H. (1985). Hyperparasitism of *Puccinia horiana* and other microcyclic rusts. *Phytopathologische Zeitschrift*, **114**, 73–8.

Strashnow, Y., Elad, Y., Sivan, A. & Chet, I. (1985). Integrated control of *Rhizoctonia solani* by methyl bromide and *Trichoderma harzianum*. *Plant Pathology*, **34**, 146–51.

Swendsrud, D.P. & Calpouzos, L. (1972). Rust uredospores increase the germination of pycnidiospores of *Darluca filum*. *Phytopathology*, **60**, 1445–7.

Traquair, J.A., Meloche, R.B., Jarvis, W.R. & Bakar, K.W. (1984). Hyperparasitism of *Puccinia violae* by *Cladosporium uredinicola*. *Canadian Journal of Botany*, **62**, 181–4.

Tronsmo, A. & Dennis, C. (1977). The use of *Trichoderma* species to control strawberry fruit rots. *Netherlands Journal of Plant Pathology*, **83** (Suppl. 1), 449–55.

Trutmann, P., Keane, P.J. & Merriman, P.R. (1980). Reduction of sclerotial inoculum of *Sclerotinia sclerotiorum* with *Coniothyrium minitans*. *Soil Biology and Biochemistry*, **12**, 461–5.

Tsuneda, A. & Skoropad, W.P. (1978a). Behaviour of *Alternaria brassicae* and its mycoparasite *Nectria inventa* on intact and on excised leaves of rapeseed. *Canadian Journal of Botany*, **56**, 1333–40.

Tsuneda, A. & Skoropad, W.P. (1978b). Nutrient leakage from dried and rewetted conidia of *Alternaria brassicae* and its effect on the mycoparasite *Nectria inventa*. *Canadian Journal of Botany*, **56**, 1341–5.

Tu, J.C. (1980). *Gliocladium virens*, a destructive mycoparasite of *Sclerotinia sclerotiorum*. *Phytopathology*, **70**, 670–4.

Tu, J.C. (1984). Mycoparasitism by *Coniothyrium minitans* on *Sclerotinia sclerotiorum* and its effect on sclerotial germination. *Phytopathologische Zeithschrift*, **109**, 261–8.

Turner, G.J. & Tribe, H.T. (1976). On *Coniothyrium minitans* and its parasitism of *Sclerotinia* species. *Transactions of the British Mycological Society*, **66**, 97–104.

Vakili, N.G. (1985). Mycoparasitic fungi associated with potential stalk rot of pathogens of corn. *Phytopathology*, **75**, 1201–7.

Velvis, H. & Jager, G. (1983). Biological control of *Rhizoctonia solani* on potatoes by antagonists. 1. Preliminary experiment with *Verticillium biguttatum*, a sclerotium-inhabiting fungus. *Netherlands Journal of Plant Pathology*, **89**, 113–23.

Weller, D.M. (1983). Colonization of wheat roots by a fluorescent pseudomonad suppressive to take-all. *Phytopathology*, **73**, 1548–53.

Weller, D.M. & Cook, R.J. (1983). Suppression of take-all of wheat by seed treatments with fluorescent pseudomonads. *Phytopathology*, **73**, 463–9.

Wells, H.D., Bell, D.K. & Jaworski, C.A. (1972). Efficacy of *Trichoderma harzianum* as biocontrol for *Sclerotium rolfsii*. *Phytopathology*, **62**, 442–7.

Wicker, E.F. (1981). Natural control of white pine blister rust by *Tuberculina maxima*. *Phytopathology*, **71**, 997–1000.

Windels, C.E. & Lindow, S.E. (eds) (1985). *Biological Control on the Phylloplane*. American Phytopathological Society, St. Paul, Minn.

Wynn, A.R. & Epton, H.A.S. (1979). Parasitism of oospores of *Phytophthora* in soil. *Transactions of the British Mycological Society*, **73**, 255–9.

Zazzerini, A. & Tosi, L. (1985). Antagonistic activity of fungi isolated from sclerotia of *Sclerotinia sclerotiorum*. *Plant Pathology*, **34**, 415–21.

Prospects for the use of fungi in nematode control

10.1 Introduction

Fungi which parasitise or prey on nematodes were first reported more than a century ago but the first attempts to utilise them as biological control agents against plant-parasitic nematodes were not undertaken until Linford and colleagues in Hawaii and Deschiens and co-workers in France began work with the nematode-trapping fungi in the late 1930s. Unfortunately for the development of biological control, interest waned following the Second World War because nematodes were being controlled cheaply and effectively with the newly developed soil fumigants. Scientific interest in plant-parasitic nematodes increased but most of the research effort went into identifying new species, understanding their ecology and host–parasite relationships and studying the movement and mode of action of nematicides. Dreschler in the USA, Duddington in England and Soprunov and colleagues in the USSR made substantial contributions to our understanding of the taxonomy and ecology of the nematode-trapping fungi but most nematologists showed no more than a casual interest in biological control.

This situation has changed in recent years as the health risks and environmental hazards associated with the use of nematicides have become apparent. There is now greater recognition of the need to develop alternatives to the chemical control strategies on which we have depended for the last 40 years. Recently, interest in fungi as biological control agents has increased and broadened to include fungi which parasitise rather than prey on nematodes and fungi parasitic in nematode eggs. This chapter reviews our knowledge of some aspects of the ecology of these fungi and examines the possibility of using them for nematode control. More detailed information on other aspects may be obtained from other recent reviews of the nematophagous fungi (Duddington & Wyborn, 1972; Barron, 1977, 1981; Cooke, 1977; Kerry, 1980, 1981, 1984, 1987; Mankau 1980a, b; Tribe 1977b, 1980).

10.2 Types of nematophagous fungi

The nematophagous fungi are so diverse that they do not fit naturally into distinct taxonomic, ecological or nutritional groups. In particular, the distinction between predaceous and endoparasitic species is not clear cut (Barron, 1977). For convenience, I have divided the nematophagous fungi into three groups, although the limitations involved in trying to categorise them in this manner are recognised.

10.2.1 *Nematode-trapping fungi*
Most students of soil biology will be familiar with the nematode-trapping fungi (Fig. 10. 1) as they have been the subject of many reviews and several popular articles over the years (Duddington, 1956, 1957, 1962; Duddington & Wyborn, 1972; Soprunov, 1966; Barron, 1977). The taxonomy of the group was revised recently (van Oorschot, 1985) and most species are found in the genera *Arthrobotrys, Dactylella, Geniculifera* and *Monacro-*

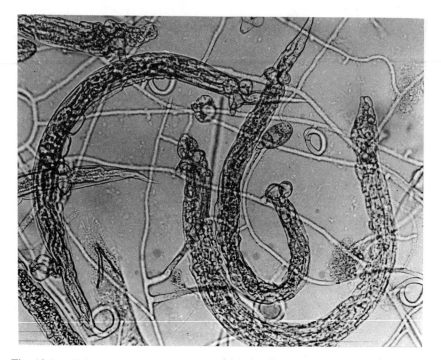

Fig. 10.1.   Nematodes trapped by a fungus with constricting rings. From Barron, G.L. (1977). *The Nematode-destroying Fungi*, page 55.

sporium. Some nematode-trapping fungi capture nematodes with an adhesive material which covers the entire surface of the hyphae, but most are characterised by the formation of specialised trapping structures on the mycelium. These well defined trapping devices are of four main types, namely adhesive networks, adhesive knobs, and non-constricting and constricting rings. The effectiveness of the traps is not simply a function of their mechanical design; the trapping organs of some species give off substances attractive to nematodes (Balan & Gerber, 1972; Field & Webster, 1977; Jansson & Nordbring-Hertz, 1979, 1980), and others produce toxins which immobilise captured nematodes or nematodes that are nearby (Olthof & Estey 1963; Balan & Gerber 1972). The nematode-trapping fungi do not depend on nematodes for their nutrition; they largely exist saprophytically in soil and therefore compete with other soil organisms for the available energy sources.

10.2.2 *Endoparasitic fungi*
There is considerable diversity within the endoparasitic fungi, and species capable of attacking plant-parasitic nematodes are found throughout the

Fig. 10.2. Scanning electron micrographs showing conidia of *Meria coniospora* attached to the mouth region (a) and tail region (b) of *Panagrellus redivivus*. Bars represent 2 μm (a) and 5 μm (b). From Jansson, H.B. and Nordbring-Hertz, B. (1983). *Journal of General Microbiology*, **129**, 1121–6.

fungal kingdom. In contrast to the nematode-trapping fungi, endoparasites do not show extensive hyphal development outside the body of the host. Most species exist in soil as adhesive spores which remain dormant until a suitable host comes into contact with them (Fig 10.2). Attached spores germinate, a germ tube penetrates the cuticle and the fungus then proliferates throughout the nematode. Growth and development occur totally within the nematode's body, although in some species conidiophores or fertile hyphae may extend from the cadaver once the body contents have been consumed. Endoparasites with adhesive spores are found in several genera including *Meria*, *Verticillium*, *Hirsutella*, *Meristacrum*, and *Cephalosporium*. In some species of *Nematoctonus*, the conidia themselves are not adhesive and the adhesive cells are borne on the germ tubes of the germinating conidia.

A group of lower fungi in genera such as *Catenaria*, *Myzocytium* and *Haptoglossa* have a somewhat different mode of infection. In the more primitive species, zoospores swim to the host and encyst on the cuticle prior to penetration and colonisation (Fig. 10.3). Since zoospores are vulnerable in the soil environment because they must locate a host before exhausting their limited food reserves, some species have evolved more sophisticated infection mechanisms which are described in detail by Barron (1977). Most of the endoparasitic lower fungi have been reported from free-living nematodes but some are adapted to attack the adult stage of sedentary endoparasitic nematodes. *Nematophthora gynophila* Kerry and Crump, *Catenaria auxiliaris* (Kuhn) Tribe and an undescribed lagenidia-

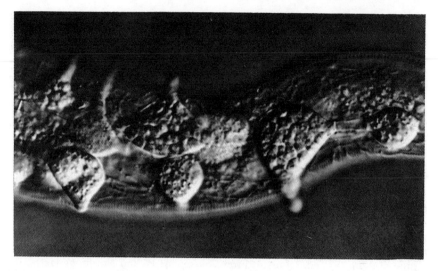

Fig. 10.3. Zoosporangia of *Catenaria anguillulae* inside a host nematode. From Barron, G.L. (1977). *The Nematode-destroying Fungi*, page 61.

ceous fungus, for example, have been reported attacking females of *Heterodera* spp. (Tribe, 1977a; Kerry & Crump, 1980). Zoospores of all three fungi encyst on females when they appear on the root surface after rupturing the cortex. The encysted spores germinate, mycelium proliferates throughout the host and then thick-walled resting spores are formed in the body of the dead nematode.

10.2.3 *Egg parasites*

A large number of fungi have been isolated from or found associated with nematode eggs (Ellis & Hesseltine, 1962; Tribe, 1977b; Stirling & Mankau, 1978a, b; Morgan-Jones *et al.*, 1981, 1983, 1984; Morgan-Jones & Rodriguez-Kabana, 1981; Gintis *et al.*, 1982; Godoy *et al.*, 1983b) and many of these are parasitic in eggs *in vitro* (Fig. 10.4). The few egg parasites that have been studied in detail form a taxonomically diverse group which includes *Verticillium chlamydosporium* Goddard, *Dactylella oviparasitica* Stirling and Mankau, *Paecilomyces lilacinus* (Thom) Samson, *Rhophalomyces elegans* Corda and *Cylindrocarpon destructans* (Zins.) Scholten. In species where the infection process has been studied, hyphae grow towards the egg and become firmly attached by means of an appressorium (Stirling & Mankau, 1979; Dunn *et al.*, 1982) or a perforation organ (Lysek, 1978). Penetration of the egg shell is probably mechanical but enzymic processes also may be involved. *D. oviparasitica*, *P. lilacinus* and several other egg parasites are known to degrade chitin (Okafor, 1967; Stirling & Mankau,

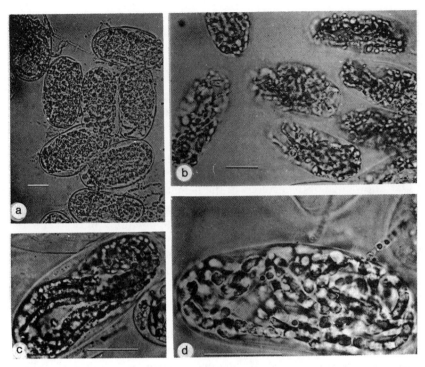

Fig. 10.4. Eggs of *Meloidogyne* sp. parasitised by *Dactylella oviparasitica*. Bars represent 25μm. From Stirling, G.R. and Mankau, R. (1978a). *Journal of Nematology*, **10**, 236–40.

1979; Godoy *et al.*, 1983a), one of the main constituents of nematode egg shells, and there appeared to be some enzymic weakening of *Meloidogyne arenaria* (Neal) Chitwood eggs parasitised by *Verticillium chlamydosporium* and *Paecilomyces lilacinus* (Morgan-Jones *et al.*, 1983, 1984).

10.3 Aspects of the ecology of nematophagous fungi

10.3.1 *Nematode-trapping fungi*
Namatode-trapping fungi have been identified from a wide range of habitats and many agricultural soils throughout the world. However, many of the habitats studied have been favourable to nematode-trapping fungi, being rich in animal manures and decaying organic matter or relatively fertile soils in moist, temperate environments (Duddington, 1954; Barron, 1977). Nematode-trapping fungi are usually isolated from 2–10 g soil samples or from larger samples from which fractions containing nematodes, fungal propagules and organic matter have been separated by decanting

and sieving. Frequency of isolation from such large samples provides no information on the number of fungal propagules in soil. A more quantitative assessment of numbers of nematode-trapping fungi can be obtained by serial dilution of soil suspensions on water agar and most probable number analysis (Eren & Pramer, 1965) but this method has rarely been used because it is time-consuming. *Arthrobotrys dactyloides* Drechsler and *Monacrosporium ellipsosporum* (Grove) Subram. were found at densities of 5–50 propagules per gram in peach orchards in California (Stirling *et al.*, 1979), but actual populations may have been higher because soil dilution methods tend to underestimate fungi which exist in soil as mycelium (Warcup, 1967).

Our knowledge of the ecology of the nematode-trapping fungi and their role in regulating nematode populations has been limited by the lack of quantitative methods of estimating fungal populations in soil and the difficulties involved in enumerating the number of nematodes killed by fungi. The literature is replete with suggestions that nematode-trapping fungi are present in large numbers in agricultural soils and may already be playing a significant role in the control of plant-parasitic nematodes, but there is little direct evidence that these fungi attack nematodes under natural conditions (Cooke, 1962b). Stirling *et al.* (1979) attempted to measure the level of predation in soil from peach orchards naturally infested with several species of nematode-trapping fungi but obtained no evidence to suggest that juveniles of *Meloidogyne incognita* (Kofoid and White) Chitwood were being trapped by these fungi.

Most of the ecological studies with nematode-trapping fungi have been carried out in soil amended with organic material. In their pioneering experiments in Hawaii, Linford *et al.* (1938) suggested that the activity of nematode-trapping fungi increased following the addition of organic matter to soil. They speculated that the increase in food supply in soil amended with organic matter led to an increase in populations of free-living nematodes and that these nematodes stimulated the activity of nematode-trapping fungi. As predation on free-living nematodes increased, plant-parasitic nematodes in the vicinity also were destroyed. This hypothesis was generally accepted until the 1960s when R.C. Cooke and others studied the nutritional requirements of the nematode-trapping fungi and tried to understand the balance between their saprophytic and predaceous phases. Cooke (1962a) used an agar disc technique to record predaceous activity in a loam soil naturally infested with several nematophagous fungi. When chopped cabbage leaf tissue was added to the soil, nematode populations increased as the organic material decomposed. The presence of a large nematode population did not result in continuous predaceous activity; the fungi trapped nematodes only for a short period after the onset of decomposition (Cooke, 1962a, b, 1963) and, 6 weeks later, predaceous activity was not detectable although nematode populations remained high. Similar experiments with soil amended with sucrose also showed that predaceous activity was not related to the size of nematode populations (Cooke, 1962b). Based on the results of these studies, Cooke (1962b, 1968) developed the view that nematode-trapping fungi in soil are in a state of low metabolic activity under normal soil conditions and predation does not occur. When organic amendments are added to soil, nematode-trapping fungi and other zymogenous fungi are stimulated into saprophytic competition for available substrates. In the early stages of decomposition there is intense competition for energy sources, and nematode-trapping fungi, being in a unique position to use an additional substrate unavailable to other soil fungi, may be triggered into predaceous activity by the nematodes present in soil. Once these energy-yielding substrates have been exhausted, predaceous activity declines. Thus predation was short-lived and was seen as a means of survival during periods of intense microbial competition and was not considered the usual mode of nutrition.

One of the problems in considering nematode-trapping fungi as a single group is that they exhibit tremendous diversity in their modes of nutrition and some are more likely to be effective biological control agents than others. Cooke (1963, 1964) studied a range of nematode-trapping fungi and found that ring-forming species and those forming adhesive knobs or branches grew poorly on corn-meal agar and had poor competitive saprophytic ability, but tended to produce traps spontaneously and to be efficient predators. In contrast, species with reticulate traps grew more rapidly in culture and had good competitive saprophytic ability, but did not produce traps spontaneously or decrease nematode populations. Therefore, some nematode-trapping fungi were more likely to prey on nematodes than others, with the more efficient predators tending to lose the capacity to compete as saprophytes in soil (Mankau, 1972).

10.3.2 *Endoparasitic fungi*

The nutritional requirements of the endoparasitic fungi can range from extreme host specificity in species such as *Nematophthora gynophila* and *Catenaria auxiliaris*, which are obligate parasites of cyst nematode females (Tribe, 1977a; Kerry & Crump, 1980), to the omnivorous food habits of *Catenaria anguillulae* Sorokin, which has been found associated with a wide range of organic materials and parasitises nematodes, rotifers, tardigrades and nematode eggs (Couch, 1945). However, most of the endoparasitic fungi appear to be largely dependent on nematodes for their nutrition, which is perhaps not unexpected considering that they produce limited mycelial growth outside the host. Some species have been cultured on standard mycological media, although they tend to grow at a very slow rate.

Soil environmental conditions affect the zoosporic fungi to a greater extent than other endoparasitic fungi. These lower fungi tend to be active only in certain soil types or during periods of the year when soil moisture is sufficient for zoospore movement (Kerry *et al.*, 1980; Stirling & Kerry, 1983) and they survive adverse conditions by producing thick-walled resting spores. Other endoparasitic fungi generally do not produce resting structures. Their spores are small and have minimal nutritional reserves, and they rely on the abundance of nematodes in soil to provide a readily available substrate (Barron, 1977). The mycelium and conidia of *Meria coniospora* Dreschler and *Cephalosporium balanoides* Dreschler are attractive to nematodes (Jansson, 1982) and this may help to ensure the continued survival of these fungi by bringing the host into close contact with the parasite.

During germination and early phases of growth, endoparasitic fungi are similar to other soil fungi and are subject to mycostasis. Conidia of *Nematoctonus concurrens* Dreschler and *N. haptocladus* Dreschler were subject to severe mycostatic and lytic effects in soil, and when germination was successful the germlings lysed (Giuma & Cooke, 1974). Spores of *Hirsutella rhossiliensis* Minter and Brady germinated in non-sterile soil but the germ tubes were short and malformed compared with those produced in sterile soil (Jaffee & Zehr, 1985). This suspectibility to mycostasis affected its competitiveness as a saprophyte. The fungus was able to colonise organic substrates such as dead nematodes or wheat seeds in sterile soil but was unable to compete with other soil fungi for these substrates in non-sterile soil. Jaffee & Zehr (1985) suggested that *H. rhossiliensis* was a specialised parasite and that it avoided antagonism by being physically separated from its competitors by the cuticle of the host nematode.

10.3.3 *Egg parasites*

There are undoubtedly many fungi capable of parasitising nematode eggs but our knowledge of their ecology is limited by the technical difficulties involved in extracting nematode eggs from soil. Most of the well-studied species parasitise eggs of nematodes such as *Heterodera* and *Meloidogyne*, where the eggs are aggregated inside cysts or in an egg mass on the root surface. The egg parasites generally appear to be excellent soil saprophytes and are not dependent on nematodes for their nutrition. For example, *Verticillium chlamydosporium* has been isolated from a variety of substrates including roots, snail eggs and fungal spores (Kerry *et al.*, 1984). *Rhophalomyces elegans* commonly occurs on rotting plant debris and animal dung (Ellis & Hesseltine, 1962), and *Paecilomyces lilacinus* is readily isolated from soil and has been found in insects and fungal sclerotia (Dunn *et al.*, 1982). The egg parasites appear to have little need to produce resting structures to ensure their survival because of the ready availability of suitable energy sources in soil. However, many isolates of *V. chlamydosporium* produce thick-walled chlamydospores under natural conditions and in culture (Kerry *et al.*, 1984).

10.4 Attempts to use fungi for nematode control

10.4.1 *Enhancement of fungi already present*

Naturally occurring fungi sometimes maintain populations of plant-parasitic nematodes below economic thresholds without being consciously manipulated by man. The best documented examples are the control of *Heterodera avenae* Woll. in European cereal fields by *Verticillium chlamydosporium* and *Nematophthora gynophila* (Kerry *et al.*, 1980, 1982a, b) and the control of root-knot nematodes in California peach orchards by *Dactylella oviparasitica* (Stirling *et al.*, 1979).

Attempts to enhance the activity of naturally occurring nematophagous fungi have generally involved the addition of organic materials to soil. Nematode populations have often been reduced and plant growth increased with organic amendments, but claims that these effects were due to the activity of fungal antagonists have never been substantiated. Naturally occurring nematode-trapping fungi have been isolated from amended soils (Linford *et al.*, 1938; Mankau & Minteer, 1962; Mankau, 1968) but their predaceous activity tends to be short-lived (Cooke, 1962a,b, 1963) and there is little direct evidence that they, or other groups of nematophagous fungi, consume large numbers of plant-parasitic nematodes. In contrast, there is good evidence that organic materials improve soil structure and plant nutrition, and numerous studies attest to the wide range of nematicidal and nematostatic compounds produced when organic materials decompose (Walker, 1971; Johnson, 1974; Sitaramaiah & Singh, 1978; Alam *et al.*, 1979; Badra *et al.*, 1979; Mian & Rodriguez-Kabana, 1982a,b,c). These factors are probably largely responsible for the beneficial effects of organic amendments.

10.4.2 *Direct addition of fungi to soil*

Attempts to achieve nematode control by introducing nematode-trapping fungi into soil have not been successful. Small-scale glasshouse and field experiments have sometimes yielded promising results (Duddington, 1962; Duddington & Wyborn, 1972), but the high level of nematode control required in modern agriculture has never been consistently achieved on a field scale. A number of commercial preparations of nematode-trapping fungi have been marketed, but these products have not been used widely because of quality control problems and inconsistent performance. *Arthrobotrys robusta* Duddington, commercially formulated as 'Royal 300'®, reduced populations of *Ditylenchus myceliophagus* Goodey and increased yields of the cultivated mushroom, *Agaricus bisporus* (Lange) Sing. (Cayrol *et al.*, 1978), while 'Royal 350'®, a similar product containing *Arthrobotrys superba* Corda, gave adequate control of root-knot nematode on tomato provided it was not used in situations where nematode populations were high (Cayrol & Frankowski, 1979; Cayrol, 1983). A commercial formulation of *A. amerospora* Schenck, Kendrick and Pramer failed to affect nematode populations, plant growth or yield in several trials in the USA (Rhoades, 1985).

Because of the difficulties involved in obtaining large numbers of propagules, the endoparasitic fungi have rarely been added to soil in attempts to control nematodes. Nematode numbers were reduced when conidia of *Nematoctonus concurrens* and *N. haptocladus* were added to sterile sand but were not affected by comparable treatments in non-sterile soil because conidia suffered severe mycostatic and lytic effects (Giuma & Cooke, 1974). These observations led the authors to question whether even massive additions of *Nematoctonus* spores to soil would result in a reduction in nematode populations. Recently, the effectiveness of *Meria coniospora* in controlling root-knot nematode on tomato was compared in sterile and non-sterile soil (Jansson *et al.*, 1985) but the value of the data is questionable because most of the nematodes used in these experiments were incapable of infecting roots.

Interest in fungi parasitising females and eggs of endoparasitic nematodes has increased rapidly in recent years, and much of this interest has been directed towards *Paecilomyces lilacinus*, which is an aggressive parasite of nematode eggs on agar (Dunn *et al.*, 1982; Morgan-Jones *et al.*, 1984). Several groups claim to have controlled root-knot and cyst nematodes in the field with *P. lilacinus* (Jatala *et al.*, 1980, 1981; Davide & Zorilla, 1983; Noe & Sasser, 1984) but, as with much of the work published on the biological control of nematodes, there was a lack of convincing evidence that the introduced fungus was responsible for the observed reductions in nematode populations. In most cases, the effects of *P. lilacinus* could not be separated from those of the substrate on which the fungus was grown; fungal populations in treated and untreated soil were never compared; and no attempt was made to assess the level of parasitic activity during the course of the experiments. Such details are important because Rodriguez-Kabana *et al.* (1984b) and Dickson & Mitchell (1985) showed that the oats or wheat used as a food base for *P. lilacinus* provided significant control of *Meloidogyne arenaria* and the addition of *P. lilacinus* did not improve the level of control. Rodriguez-Kabana *et al.* (1984b) were unable to isolate the fungus from root-knot nematodes at the end of their experiments.

Recently, investigations began into the possibility of introducing *Verticillium chlamydosporium* into soil to control *Heterodera avenae* (Kerry *et al.*, 1984). The fungus was grown on agar, on attapulgite clay granules soaked in potato dextrose broth, or on ground oat grains, and the largest reductions in nematode populations were obtained when oat grains were used as inoculum. Possibly the food base provided by the substrate enabled the fungus to overcome fungistasis and gave it a competitive advantage over resident microflora.

10.5 Aspects requiring more attention

10.5.1 *The search for better antagonists*
Many nematophagous fungi were first discovered by plating small quantities of soil or organic matter on to a weak nutrient medium in petri dishes. Although such indirect methods have proved invaluable, they tend to favour aggressive predaceous fungi and endoparasites of free-living nematodes. More useful fungi may be found by pursuing species that are actually attacking plant-parasitic nematodes in soil. This can be done by looking for parasites in females and eggs of sedentary endoparasitic nematodes, by examining moribund nematodes extracted from soil using sugar centrifugation techniques or by adding nematodes or their eggs to soil and observing them later for signs of parasitism. The recent discovery of several new fungal parasites of females and eggs of root-knot and cyst nematodes (Tribe, 1977a; Stirling & Mankau, 1978a; Jatala *et al.*, 1979; Kerry & Crump, 1980) shows that antagonists of particular plant-parasitic nematodes can be found if they are sought in a directed manner. Nematophagous fungi may be found in any random soil sample, but they should be specifically sought in areas where nematode damage is minimal, where numbers of a particular plant-parasitic nematode have declined or where high nematode populations do not develop despite the presence of a suitable soil environment and a susceptible host.

In the past, most of the fungi considered to have potential as biological control agents have been parasitic or predaceous on nematodes. Other fungal groups should not be neglected in the search for better antagonists. Many different types of interaction between vesicular–arbuscular mycorrhizae and nematodes have been observed and it may be possible to use these fungi to offset the damage caused by nematodes (Hussey & Roncadori, 1982). Fungi which produce nematicidal compounds *in vitro* (Mankau, 1969; Giuma & Cooke, 1971, 1973; Alam *et al.*, 1973; Mani & Sethi, 1984; Jatala *et al.*, 1985) may warrant further attention, particularly if it can be demonstrated that these chemicals are produced in nematicidal concentrations in soil.

10.5.2 *Selecting and evaluating fungi for nematode control*
Most of the important plant-parasitic nematodes are endoparasites and for part of their life cycles their parasitic stages and eggs are protected from

antagonists in soil. Also, their free-living stages tend to be transient and may only be susceptible to parasites and predators for limited periods. The process of selecting antagonists with potential as biological control agents should therefore begin by identifying those stages in the life cycle of the target nematode most vulnerable to antagonism. This aspect has received insufficient attention in the past and attempts to achieve biological control have been made with fungi quite unsuited to the task. For example, fungi active against second-stage juveniles of *Heterodera* and *Globodera* were used almost exclusively in biological control experiments until the early 1970s, when it was suggested that fungi which destroyed eggs were more likely to be effective antagonists (Willcox & Tribe, 1974).

A preliminary assessment of the potential of a fungus as a biological control agent can be obtained by assessing predaceous or parasitic activity *in vitro* (Heintz, 1978) or in sterilised soil in pots. However, there is little convincing evidence that antagonistic activity in such simple systems is correlated with effects in the more complex soil environment. It is essential that additional testing be carried out in natural soils where an added fungus may suffer fungistasis, competition or mycoparasitism. Ideally, closely monitored laboratory and glasshouse experiments should be conducted to determine the fate of the fungus after its introduction into field soil and to identify the factors that influence its establishment, level of antagonistic activity and survival. Without such information, field trials with introduced fungi have little chance of success.

10.5.3 *Problems of establishment*

The fungistatic nature of soil must be considered when attempts are made to establish alien fungi in soil or to activate naturally occurring fungi. In fact, it has been suggested that fungistatic effects in natural soils are so severe that they are likely to prevent the successful use of both nematode-trapping and endoparasitic fungi as biological control agents (Cooke & Satchuthananthavale, 1968; Giuma & Cooke, 1974). However, the possibility that fungistatic effects on nematophagous fungi could be reduced or circumvented has rarely been explored and is a subject worthy of further research. Nematologists are now in the position of being able to benefit from the work done with fungi introduced into soil for the control of soil-borne diseases. In recent years, plant pathologists have attempted to overcome fungistasis by improving formulation methods and selecting the fungal strains and propagule types most likely to establish in soil. Ultra-violet irradiation has been used to induce mutant strains of *Trichoderma harzianum* Rifai with enhanced ability to survive (Papavizas *et al.*, 1982). A range of materials, including diatomaceous earth impregnated with molasses, wheat-bran preparations, perlite impregnated with molasses and methyl cellulose, and lignite–stillage granules have proved suitable carriers for some biocontrol fungi (Backman & Rodriguez-Kabana, 1975; Elad *et al.*, 1980; Papavizas & Lewis, 1981; Jones *et al.*, 1984). Fungi such as *Trichoderma*, whose conidia are susceptible to fungistasis and fail to germinate in soil, have been established in natural soils by incubating conidia in sterile moist bran fro 1–3 days and using the germinated spores and young mycelium as inoculum (Lewis & Papavizas, 1984). Apparently hyphae occupying a food base were less susceptible than conidia to fungistasis. There have been suggestions that establishment of introduced fungi might be improved by using pesticides or treatments such as solarisation to eliminate major groups of soil microorganisms. Although such treatments create only a partial and temporary biological vacuum in soil, they may be sufficient to allow an introduced fungus to germinate and proliferate (Papavizas, 1985). Since soil fungistasis can be annulled by the addition of energy-containing nutrients to soil (Mankau, 1962; Lockwood, 1977), it may also be possible to stimulate spore germination and growth of introduced nematophagous fungi by the judicious use of soil amendments.

The problem of establishing a nematophagous fungus in a new environment is compounded by the fact that it must colonise and remain active in the rhizosphere, the most microbially active zone in soil. *Arthrobotrys oligospora* Fres. and *Dactylella oviparasitica* occupy the rhizosphere in preference to the general soil mass (Peterson & Katznelson, 1965; Stirling in *et al.*, 1979), and *Verticillium chlamydosporium* has been observed in the rhizosphere after its introduction to soil (Kerry *et al.*, 1984). However, more information is needed on the capacity of other species to establish and survive in this zone.

10.5.4 *Density dependence*
To successfully regulate a pest population, the proportion of individuals killed by an antagonist must increase as the density of the pest increases (Knipling, 1979). This phenomenon has been extensively investigated for parasites and predators of insects (Varley *et al.*, 1973,) but it is rarely mentioned in literature on fungal antagonists of nematodes. Since the occurrence of facultative parasites and predators of nematodes is determined by the availability of food sources other than nematodes, their level of activity is unlikely to be related to nematode density. In contrast, obligate parasites such as *Nematophthora gynophila* exhibit density dependence (Kerry *et al.*, 1982a).

One of the deficiencies of host-specific antagonists is that they tend to be ineffective biological control agents when the host density is low. For example, fungal parasites of *Heterodera avenae* provide adequate nematode control in England, where nematode populations and economic thresholds are high, but appear to be less effective in Australia where nemaotde populations and economic thresholds are much lower (Kerry *et al.*, 1982a;

Stirling & Kerry, 1983). An ideal biological control agent would be effective at the low nematode densities usually present when a crop is planted, and would also kill a large and increasing proportion of the population as nematodes multiplied during the growing season. Although a fungus with these capacities probably does not exist, the possibility of achieving a similar effect using a combination of density-dependent and density-independent antagonists, or by integrating the use of fungi with other control measures, should be investigated.

10.5.5 *The use of amendments*

When organic amendments are added to soil at more than 1% w/w (approximately 20 t/ha incorporated to a depth of 15 cm), nematicidal compounds such as ammonia are produced during the decomposition cycle (Walker, 1971; Mian *et al.*, 1982; Mian & Rodriguez-Kabana, 1982a, b; Heubner *et al.*, 1983). The nematicidal effect of large quantities of organic matter can be utilised in situations where it is cheap and readily available or where high-value crops are involved, but is impractical or too expensive to use in broad-scale agriculture. A reduction in the amount of organic matter required to obtain nematode control is needed, and this is most likely to be achieved by utilising specific amendments to stimulate and sustain a microflora antagonistic to nematodes.

Complex changes in the microbiology of soil take place when an antagonist is introduced on an organic substrate or when an amendment is used to stimulate naturally occurring antagonists. These changes must be understood if organic amendments are to aid in the establishment of useful organisms in soil. Application rates and times will need to be adjusted so that peaks of antagonistic activity correspond to periods when susceptible stages of the nematode are present in the field. Materials of plant origin have traditionally been used as amendments but more consideration should be given to the types of amendment likely to be useful. Chitinous waste from the seafood processing industry has been used to stimulate an adaptive microflora consisting largely of chitinolytic fungi (Mian *et al.*, 1982; Godoy *et al.*, 1983a; Rodriguez-Kabana *et al.*, 1983, 1984a). Many of these fungi were able to parasitise eggs of root-knot and cyst nematodes, whose egg shells contain chitin, suggesting that it may be possible to add specific amendments to soil and develop a specialised mycoflora which is antagonistic towards nematodes.

Our capacity to manipulate the soil microflora leads to the question of whether the future of biological control lies with the introduction of alien fungi or with the encouragement of naturally occurring species. A typical fertile agricultural soil contains bacteria, fungi, actinomycetes and protozoans at populations ranging from 10^4 to 10^9 propagules per gram, plus numerous nematodes, tardigrades, insects, mites and other small animals, a habitat so complex that alien fungi cannot be established by simply adding spores or mycelium to soil. Although a substrate added with the fungus provides a food base from which colonisation can proceed, the introduced fungus still suffers competition from other soil organisms. Because of the ubiquitous distribution of many nematophagous fungi, we may have more success if amendments such as chitin were used to stimulate the activity of naturally occurring, opportunistic fungi capable of also utilising or affecting plant-parasitic nematodes. However, there are limitations to the potential usefulness of such strategies. They may be of limited value in deep soils or with deep-rooted crops because of the difficulties involved in incorporating amendments and stimulating fungal activity in lower soil layers. When *Paecilomyces lilacinus* was introduced into Florida soils, for example, the fungus was not recovered from depths below 15 cm (D. Dickson, personal communication). Also, it is unlikely that such strategies could be used in the same way as nematicides to provide immediate nematode control. They should be seen as essentially a long-term solution to nematode problems because multiple applications and time would be needed to establish an antagonistic microflora in soil.

10.5.6 *Hazards to non-target species*
Some species of *Paecilomyces*, including *P. lilacinus*, cause eye infections and facial lesions in humans and infections in domestic animals (Chandler *et al.*, 1980). Isolates of *P. lilacinus* from nematode eggs and soil have yet to be implicated in such diseases, but extensive medical tests will be needed to resolve this matter. Fungi such as *Aspergillus*, *Verticillium* and *Fusarium* are often associated with or parasitic in eggs of plant-parasitic nematodes (Morgan-Jones & Rodriguez-Kabana, 1981; Morgan-Jones *et al.*, 1981; Gintis *et al.*, 1982; Godoy *et al.*, 1983b) and species in these genera also cause diseases of plants and man. Even if medical, veterinary or plant pathological studies showed that such fungi acted specifically against nematodes, some registration authorities may be unwilling to sanction their commercial use because of their close relationship with known pathogens.

10.5.7 *Experimental techniques*
If we are to control plant-parasitic nematodes by introducing nematophagous fungi or manipulating naturally occurring fungal antagonists, we must increase our understanding of the nutritional requirements, ecology and mode of action of these fungi. In particular, the role of organic amendments in stimulating nematophagous fungi must be investigated more thoroughly. Unfortunately, much of the experimental work published in recent years fails to address these issues. The typical field experiment, in which a fungus–substrate preparation is applied and nematode populations and plant growth parameters are measured at the end of the season, is of

limited value. Treatments such as the fungus alone, the substrate alone and autoclaved colonised substrate should be included as standard practice, particularly when an organism is being tested for the first time. Although the effects of such treatments may be difficult to interpret (Baker *et al.*, 1984), their inclusion creates awareness of the multiplicity of factors acting in such experiments and should lead to attempts to understand the interactions involved. Claims that a fungus has acted as a parasite, predator or toxin producer in field or glasshouse experiments should be supported by evidence that such activity has been observed and measured. In situations where direct methods cannot be used, it should not be difficult to develop appropriate laboratory bioassays in which nematodes or their eggs are added to soil and later recovered and examined for the effects of antagonists. Careful attention should be given to whether a fungus is actually parasitic, or simply living saprophytically on dead nematodes or their eggs. *Catenaria anguillulae*, for example, is largely saprophytic in dead or injured nematodes or parasitic in nematodes stressed by sub-lethal doses of nematicides (Boosalis & Mankau, 1965; Roy, 1982). The decision to commence field trials should be made carefully because potentially useful organisms are often tested in the field when appropriate laboratory studies might yield more information. R.C. Cooke's experiments with the nematode-trapping fungi (Cooke 1962a,b, 1963, 1964) and the studies of Rodriguez-Kabana and colleagues with organic amendments (Mian & Rodriguez-Kabana, 1982a,b,c; Mian *et al.*, 1982; Godoy *et al.*, 1983a; Heubner *et al.*, 1983; Rodriguez-Kabana *et al.*, 1984a) provide a good example of how our basic understanding of the processes of biological control can be improved through a series of simple laboratory and glasshouse studies.

10.6 Future prospects

Modern agriculture demands nematode management strategies that are reliable, effective and economically viable and which can be satisfactorily integrated with overall production systems. Many of those interested in the possibility of using fungi as biological control agents have tended to underestimate the difficulties involved in satisfying these criteria. Plant-parasitic nematodes live in a complex environment where fungi are subject to competition, antibiosis, lysis and mycoparasitism. Endoparasitic nematodes live in roots for much of their life cycle and are therefore protected from fungal antagonists in the soil and rhizosphere. Large quantities of substrate or organic additives generally have been required to introduce fungi to soil or to modify the soil environment, making such practices either impractical or prohibitively expensive (Kerry, 1984). Finally, nematophagous fungi must survive and remain active in production systems where cultivation practices, crop rotation schemes and nematicide and fungicide usage are likely to be deleterious.

Biological control with nematophagous fungi will obviously not be achieved easily; fifty years of failed attempts attest to that fact. However, some recent observations provide cause for optimism. We now have evidence that naturally occurring fungi sometimes keep nematode populations under control, our understanding of the effects of organic amendments on the soil microflora is improving, and several promising new parasites of root-knot and cyst nematodes have been found. Nevertheless, further in-depth ecological studies are required if fungi are to become a reliable and practical means of nematode control.

References

Alam, M.M., Khan, W.M. & Saxena, S.K. (1973). Inhibiting effect of culture filtrates of some rhizosphere fungi of okra on the mortality and larval hatch of certain plant parasitic nematodes. *Indian Journal of Nematology*, 3, 94–8.

Alam, M.M., Khan, A.M. & Saxena, S.K. (1979). Mechanism of control of plant parasitic nematodes as a result of the application of organic amendments to the soil. V. — Role of phenolic compounds. *Indian Journal of Nematology*, 9, 136–42.

Backman, P.A. & Rodriguez-Kabana, R. (1975). A system for the growth and delivery of biological control agents to soil. *Phytopathology*, 65, 819–21.

Badra, T., Saleh, M.A. & Oteifa, B.A. (1979). Nematicidal activity and composition of some organic fertilizers and amendments. *Revue de Nematologie*, 2, 29–36.

Baker, R., Elad, Y. & Chet, I. (1984). The controlled experiment in the scientific method with special emphasis on biological control. *Phytopathology*, 74, 1019–21.

Balan, J. & Gerber. N. (1972). Attraction and killing of the nematode *Panagrellus redivivus* by the predacious fungus *Arthrobotrys dactyloides*. *Nematologica*, 18, 163–73.

Barron, G.L. (1977). *The Nematode-destroying Fungi*. Canadian Biological Publications, Guelph.

Barron, G.L. (1981). Predators and parasites of microscopic animals. In: *Biology of Conidial Fungi, Vol. 1*, eds G.T. Cole & B. Kendrick, pp. 167–200. Academic Press, New York.

Boosalis, M.G. & Mankau, R. (1965). Parasitism and predation of soil microorganisms. In: *Ecology of Soil-borne Plant pathogens*, eds K.F. Baker and W.C. Snyder, pp. 374–91, University of California Press, Berkeley, CA.

Cayrol, J.C. (1983). Lutte biologique contre les *Meloidogyne* au moyen d'*Arthrobotrys irregularis*. *Revue de Nematologie*, 6, 265–73.

Cayrol, J.C. & Frankowski, J.P. (1979). Une méthode de lutte biologique contre les nématodes à galles des racines appartenant au genre *Meloidogyne*. *Pépiniéristes, Horticulteurs, Maraîchers–Revue Horticole*, 193, 15–23.

Cayrol, J.C., Frankowski, J.P., Laniece, A., D'Hardemare, G. & Talon, J.P. (1978). Contre les nématodes en champignonnière. Mise au point d'une methode de lutte biologique á l'aide d'un Hyphomycete predateur: *Arthrobotrys robusta* souche 'antipolis' (Royal 300). *Pépiniéristes, Horticulteurs, Maraîchers — Revue Horticole*, 184, 23–30.

Chandler, F.W., Kaplan, W. & Ajello, L. (1980). *A Colour Atlas and Textbook of the Histopathology of Mycotic Diseases*. Wolfe Medical Publications, London.

Cooke, R.C. (1962a). Behaviour of nematode-trapping fungi during decomposition of organic matter in soil. *Transactions of the British Mycological Society*, **45**, 314–20.

Cooke, R.C. (1962b). The ecology of nematode-trapping fungi in the soil. *Annals of Applied Biology*, **50**, 507–13.

Cooke, R.C. (1963). Succession of nematophagous fungi during the decomposition of organic matter in the soil. *Nature, London*, **197**, 205.

Cooke, R.C. (1964). Ecological characteristics of nematode-trapping hyphomycetes. II. Germination of conidia in soil. *Annals of Applied Biology*, **54**, 375–9.

Cooke, R.C. (1968). Relationships between nematode-destroying fungi and soil-borne phytonematodes. *Phytopathology*, **58**, 909–13.

Cooke, R.C. (1977). *The Biology of Symbiotic Fungi*. John Wiley, Chichester.

Cooke, R.C. & Satchuthananthavale, V. (1968). Sensitivity to mycostasis of nematode-trapping hyphomycetes. *Transactions of the British Mycological Society*, **51**, 555–61.

Couch, J.N. (1945). Observations on the genus *Catenaria*. *Mycologia*, **37**, 163–93.

Davide, R.G. & Zorilla, R.A. (1983). Evaluation of a fungus, *Paecilomyces lilacinus* (Thom.) Samson, for the biological control of the potato cyst nematode, *Globodera rostochiensis* Woll. as compared with some nematicides. *Philippines Agriculturist*, **66**, 397–404.

Dickson, D.W. & Mitchell, D.J. (1985). Evaluation of *Paecilomyces lilacinus* as a biological control agent of *Meloidogyne javanica* on tobacco. *Journal of Nematology*, **17**, 519.

Duddington, C.L. (1954). Nematode-destroying fungi in agricultural soils. *Nature, London*, **168**, 38–9.

Duddington, C.L. (1956). The predacious fungi: Zoopagales and Moniliales. *Biological Reviews*, **31**, 152–93.

Duddington, C.L. (1957). *The Friendly Fungi*. Faber & Faber, London.

Duddington, C.L. (1962). Predacious fungi and the control of eelworms. In: *Viewpoints in Biology, Vol. 1*, eds C.L. Duddington & J.D. Carthy, pp. 151–200. Butterworths, London.

Duddington, C.L. & Wyborn, C.H.E. (1972). Recent research on the nematophagous Hyphomycetes. *Botanical Review*, **38**, 545–65.

Dunn, M.T., Sayre, R.M., Carrell, A. & Wergin, W.P. (1982). Colonization of nematode eggs by *Paecilomyces lilacinus* (Thom.) Sampson as observed with scanning electron microscope. *Scanning Electron Microscopy*, **3**, 1351–7.

Elad, Y.I., Chet, I. & Katan, J. (1980). *Trichoderma harzianum*: biocontrol agent effective against *Sclerotium rolfsii* and *Rhizoctonia solani*. *Phytopathology*, **70**, 119–21.

Ellis, J.J. & Hesseltine, C.W. (1962). *Rhophalomyces* and *Spinellus* in pure culture and the parasitism of *Rhophalomyces* on nematode eggs. *Nature, London*, **193**, 699–700.

Eren, J. & Pramer, D. (1965). The most probable number of nematode trapping fungi in soil. *Soil Science*, **99**, 285.

Field, J.I. & Webster, J. (1977). Traps of predacious fungi attract nematodes. *Transactions of the British Mycological Society*, **68**, 467–9.

Gintis, B.O., Morgan-Jones, G. & Rodriguez-Kabana, R. (1982). Mycoflora of young cysts of *Heterodera glycines* in North Carolina soils. *Nematropica*, **12**, 295–303.

Giuma, A.Y. & Cooke, R.C. (1971). Nematoxin production of *Nematoctonus haptocladus* and *N. concurrens*. *Transactions of the British Mycological Society*, **56**, 89–94.

Giuma, A.Y. & Cooke, R.C. (1973). Thermostable nematoxins produced by germinating conidia of some endozoic fungi. *Transactions of the British Mycological Society*, **60**, 49–56.

Giuma, A.Y. & Cooke, R.C. (1974). Potential of *Nematoctonus* conidia for biological control of soil-borne phytonematodes. *Soil Biology and Biochemistry*, **6**, 217–20.

Godoy, G., Rodriguez-Kabana, R., Shelby, R.A. & Morgan-Jones, G. (1983a). Chitin amendments for control of *Meloidogyne arenaria* in infested soil. II. Effects on microbial population. *Nematropica*, **13**, 63–74.

Godoy, G., Rodriguez-Kabana, R. & Morgan-Jones, G. (1983b). Fungal parasites of *Meloidogyne arenaria* eggs in an Alabama soil. A mycological survey and greenhouse studies. *Nematropica*, **13**, 201–13.

Heintz, C.E. (1978). Assessing the predacity of nematode-trapping fungi *in vitro*. *Mycologia*, **70**, 1086–1100.

Heubner, R.A., Rodriguez-Kabana, R. & Patterson, R.M. (1983). Hemicellulosic waste and urea for control of plant-parasitic nematodes: effect on soil enzyme activities. *Nematropica*, **13**, 37–54.

Hussey, R.S. & Roncadori, R.W. (1982). Vesicular–arbuscular mycorrhizae may limit nematode activity and improve plant growth. *Plant Disease*, **66**, 9–14.

Jaffee, B.A. & Zehr, E.I. (1985). Parasitic and saprophytic abilities of the nematode-attacking fungus *Hirsutella rhossiliensis*. *Journal of Nematology*, **17**, 341–5.

Jansson, H.-B. (1982). Attraction of nematodes to endoparasitic nematophagous fungi. *Transactions of the British Mycological Society*, **79**, 25–9.

Jansson, H.-B. & Nordbring-Hertz, B. (1979). Attraction of nematodes to living mycelium of nematophagous fungi. *Journal of General Microbiology*, **112**, 89–93.

Jansson, H.-B & Nordbring-Hertz, B. (1980). Interactions between nematophagous fungi and plant-parasitic nematodes: attraction, induction of trap formation and capture. *Nematologica*, **26**, 383–9.

Jansson, H.-B., Jeyaprakash, A. & Zuckerman, B.M. (1985). Control of root-knot nematodes on tomato by the endoparasitic fungus *Meria coniospora*. *Journal of Nematology*, **17**, 327–9.

Jatala, P., Kaltenbach, R. & Bocangel, M. (1979). Biological control of *Meloidogyne incognita acrita* and *Globodera pallida* on potatoes. *Journal Nematology*, **11**, 303.

Jatala, P., Franco, J., Gonzalez, A. & O'Hara, C.M. (1985). Hatching stimulation and inhibition of *Globodera pallida* eggs by the enzymatic and exopathic toxic compounds of some biocontrol fungi. *Journal of Nematology*, **17**, 501.

Jatala, P., Kaltenbach, R., Bocangel, M., Devaux, A.J. & Campos, R. (1980). Field application of *Paecilomyces lilacinus* for controlling *Meloidogyne incognita* on potatoes. *Journal of Nematology*, **12**, 226–7.

Jatala, P., Sales, R., Kaltenbach, R. & Bocangel, M. (1981). Multiple application and long-term effect of *Paecilomyces lilacinus* in controlling *Meloidogyne incognita* under field conditions. *Journal of Nematology*, **13**, 445.

Johnson, L.F. (1974). Extraction of oat straw, flax and amended soil to detect substances toxic to root-knot nematode. *Phytopathology*, **64**, 1471–3.

Jones, R.W., Pettit, R.E. & Taber, R.A. (1984). Lignite and stillage: carrier and substrate for application of fungal biocontrol agents to soil. *Phytopathology*, **74**, 1167–70.

Kerry, B.R. (1980). Biocontrol: fungal parasites of female cyst nematodes. *Journal of Nematology*, **12**, 253–9.

Kerry, B.R. (1981). Fungal parasites: a weapon against cyst nematodes. *Plant Disease*, **65**, 390–3.

Kerry, B.R. (1984). Nematophagous fungi and the regulation of nematode populations in soil. *Helminthological Abstracts Series B*, **53**, 1–14.

Kerry, B.R. (1987). Biological control. In: *Principles and Practice of Nematode Control in Crops*, eds R.H. Brown & B.R. Kerry, pp. 233–63. Academic Press, Sydney.

Kerry, B.R. & Crump, D.H. (1980). Two fungi parasitic on females of cyst nematodes (*Heterodera* spp.). *Transactions of the British Mycological Society*, **74**, 119 –25.

Kerry, B.R., Crump, D.H. & Mullen, L.A. (1980). Parasitic fungi, soil moisture and multiplication of the cereal cyst nematode, *Heterodera avenae*. *Nematologica*, **26**, 57–68.

Kerry, B.R., Crump, D.H. & Mullen, L.A. (1982a). Studies of the cereal cyst-nematode, *Heterodera avenae*, under continuous cereals, 1975–1978. II. Fungal parasitism of nematode females and eggs. *Annals of Applied Biology*, **100**, 489 –99.

Kerry, B.R., Crump, D.H. & Mullen, L.A. (1982b). Natural control of the cereal cyst-nematode, *Heterodera avenae* Woll., by soil fungi at three sites. *Crop Protection*, **1**, 99–109.

Kerry, B.R., Simon, A. & Rovira, A.D. (1984). Observations on the introduction of *Verticillium chlamydosporium* and other parasitic fungi into soil for control of the cereal cyst-nematode *Heterodera avenae*. *Annals of Applied Biology*, **105**, 509–16.

Knipling, E.F. (1979). The basic principles of insect population suppression and management. *USDA Agriculture Handbook No. 512*. US Government Printing Office, Washington, DC.

Lewis, J.A. & Papavizas, G.C. (1984). A new approach to stimulate population proliferation of *Trichoderma* species and other potential biocontrol fungi introduced into natural soils. *Phytopathology*, **74**, 1240–4.

Linford, M.B., Yap, F. & Oliveira, J.M. (1938). Reduction of soil populations of the root-knot nematode during decomposition of organic matter. *Soil Science*, **45**, 127–41.

Lockwood, J.L. (1977). Fungistasis in soils. *Biological Reviews*, **52**, 1–43.

Lysek, H. (1978). A scanning electron microscope study of the effect of an ovicidal fungus on the eggs of *Ascaris lumbricoides*. *Parasitology*, **77**, 139–41.

Mani, A. & Sethi, C.L. (1984). Effect of culture filtrates of *Fusarium oxysporum* f.sp. *ciceri* and *Fusarium solani* on hatching and juvenile mobility of *Meloidogyne javanica*. *Nematropica*, **14**, 139–44.

Mankau, R. (1962). Soil fungistasis and nematophagous fungi. *Phytopathology*, **52**, 611–15.

Mankau, R. (1968). Reduction of root-knot disease with organic amendments under semi-field conditions. *Plant Disease Reporter*, **52**, 315–19.

Mankau, R. (1969). Nematicidal activity of *Aspergillus niger* culture filtrates. *Phytopathology*, **59**, 1170.

Mankau, R. (1972). Utilization of parasites and predators in nematode pest management ecology. *Proceedings, Tall Timbers Conference on Ecological Animal Control by Habitat Management*, **4**, 129–43.

Mankau, R. (1980a). Biocontrol: fungi as nematode control agents. *Journal of Nematology*, **12**, 244–52.

Mankau, R. (1980b). Biological control of nematode pests by natural enemies.

Annual Review of Phytopathology, **18**, 415–40.

Mankau, R. & Minteer, R.J. (1962). Reduction of soil populations of the citrus nematode by the addition of organic materials. *Plant Disease Reporter*, **46**, 375–8.

Mian, I.H., Godoy, G., Shelby, R.A., Rodriguez-Kabana, R. & Morgan-Jones, G. (1982). Chitin amendments for control of *Meloidogyne arenaria* in infested soil. *Nematropica*, **12**, 71–84.

Mian, I.H. & Rodriguez-Kabana, R. (1982a). Survey of the nematicidal properties of some organic materials available in Alabama as amendments to soil for control of *Meloidogyne arenaria*. *Nematropica*, **12**, 235–46.

Main, I.H. & Rodriguez-Kabana, R. (1982b). Soil amendments with oilcakes and chicken litter for control of *Meloidogyne arenaria*. *Nematropica*, **12**, 205–20.

Mian, I.H. & Rodriguez-Kabana, R. (1982c). Organic amendments with high tannin and phenolic contents for control of *Meloidogyne arenaria* in infested soil. *Nematropica*, **12**, 221–34.

Morgan-Jones, G., Gintis, B.O. & Rodriguez-Kabana, R. (1981). Fungal colonisation of *Heterodera glycines* cysts in Arkansas, Florida, Mississippi and Missouri soils. *Nematropica*, **11**, 155–63.

Morgan-Jones, G. & Rodriguez-Kabana, R. (1981). Fungi associated with cysts of *Heterodera glycines* in an Alabama soil. *Nematropica*, **11**, 69–74.

Morgan-Jones, G., White, J.F. & Rodriguez-Kabana, R. (1983). Phytonematode pathology: ultrastructural studies. I. Parasitism of *Meloidogyne arenaria* eggs by *Verticillium chlamydosporium*. *Nematropica*, **13**, 245–60.

Morgan-Jones, G., White, J.F. and Rodriguez-Kabana, R. (1984). Phytonematode pathology: ultrastructural studies. II. Parasitism of *Meloidogyne arenaria* eggs and larvae by *Paecilomyces lilacinus*. *Nematropica*, **14**, 57–71.

Noe, J.P. & Sasser, J.N. (1984). Efficacy of *Paecilomyces lilacinus* in reducing yield losses due to *Meloidogyne incognita*. *Proceedings of the First International Congress of Nematology*, 61.

Okafor, N. (1967). Decomposition of chitin by micro-organisms isolated from a temperate soil and a tropical soil. *Nova Hedwigia*, **13**, 209–26.

Olthof, T.H.A. & Estey, R.H. (1963). A nematoxin produced by the nematophagous fungus *Arthrobotrys oligospora* Fresenius. *Nature, London*, **197**, 514–15.

Papavizas, G.C. (1985). *Trichoderma* and *Gliocladium*: biology, ecology and potential for biocontrol. *Annual Review of Phytopathology*, **23**, 23–54.

Papavizas, G.C. & Lewis, J.A. (1981). Introduction and augmentation of microbial antagonists for the control of soilborne plant pathogens. In: *Biological Control in Crop Production*, ed. G.C. Papavizas, pp. 305–22. Allanheld & Osmun, Totowa, N.J.

Papavizas, G.C., Lewis, J.A. & Abd El-Moity, T.H. (1982). Evaluation of new biotypes of *Trichoderma harzianum* for tolerance to benomyl and enhanced biocontrol capabilities. *Phytopathology*, **72**, 126–32.

Peterson, E.A. & Katznelson, H. (1965). Studies on the relationships between nematodes and other soil microorganisms. IV. Incidence of nematode-trapping fungi in the vicinity of plant roots. *Canadian Journal of Microbiology*, **11**, 491–5.

Rhoades, H.L. (1985). Comparison of fenamiphos and *Arthrobotrys amerospora* for controlling plant nematodes in central Florida. *Nematropica*, **15**, 1–7.

Rodriguez-Kabana, R., Godoy, G., Morgan-Jones, G. & Shelby, R.A. (1983). The determination of soil chitinase activity: conditions for assay and ecological studies. *Plant and Soil*, **75**, 95–106.

Rodriguez-Kabana, R., Morgan-Jones, G. & Gintis, B.O. (1984a). Effects of chitin amendments to soil on *Heterodera glycines*, microbial populations and colonization of cysts by fungi. *Nematropica*, **14**, 10–25.

Rodriguez-Kabana, R., Morgan-Jones, G., Godoy, G. & Ginits, B.O. (1984b). Effectiveness of species of *Gliocladium, Paecilomyces* and *Verticillium* for control of *Meloidogyne arenaria* in field soil. *Nematropica*, **14**, 155–70.

Roy, A.K. (1982). Effect of ethoprop on the parasitism of *Catenaria anguillulae* on *Meloidogyne incognita*. *Revue de Nematologie*, **5**, 335–6.

Sitaramaiah, K. & Singh, R.S. (1978). Role of fatty acids in margora cake applied as soil amendment in the control of nematodes. *Indian Journal of Agricultural Science*, **48**, 266–70.

Soprunov, F.F. (1966). *Predacious Hyphomycetes and their Application in the Control of Pathogenic Nematodes*. Israel Program for Scientific Translations, Jerusalem.

Stirling, G.R. & Kerry, B.R. (1983). Antagonists of the cereal cyst nematode, *Heterodera avenae* Woll. in Australian soils. *Australian Journal of Experimental Agriculture and Animal Husbandry*, **23**, 318–24.

Stirling, G.R. & Mankau, R. (1978a). *Dactylella oviparasitica*, a new fungal parasite of *Meloidogyne* eggs. *Mycologia*, **70**, 774–83.

Stirling, G.R. & Mankau, R. (1978b). Parasitism of *Meloidogyne* eggs by a new fungal parasite. *Journal of Nematology*, **10**, 236–40.

Stirling, G.R. & Mankau, R. (1979). Mode of parasitism of *Meloidogyne* and other nematode eggs by *Dactylella oviparasitica*. *Journal of Nematology*, **11**, 282–8.

Stirling, G.R. McKenry, M.V. & Mankau, R. (1979). Biological control of root-knot nematodes (*Meloidogyne* spp.) on peach. *Phytopathology*, **69**, 806–9.

Tribe, H.T. (1977a). A parasite of white cysts of *Heterodera*. *Catenaria auxiliaris* (Kuhn) comb. nov. *Transactions of the British Mycological Society*, **69**, 367–76.

Tribe, H.T. (1977b). Pathology of cyst nematodes. *Biological Reviews*, **52**, 477–507.

Tribe, H.T. (1980). Prospects for the biological control of plant-parasitic nematodes. *Parasitology*, **81**, 619–39.

Van Oorschot, C.A.N. (1985). Taxonomy of the *Dactylaria* complex, V. A review of *Arthrobotrys* and allied genera. *Studies in Mycology*, **26**, 61–96.

Varley, C.G., Gradwell, G.R. & Hassell, M.P. (1973). *Insect Population Ecology*. University of California Press, Berkeley, CA.

Walker, J.T. (1971). Populations of *Pratylenchus penetrans* relative to decomposing nitrogenous soil amendments. *Journal of Nematology*, **3**, 43–9.

Warcup, J.H. (1967). Fungi in soil. In: *Soil Biology*, eds A. Burges and F. Raw, pp. 50–110. Academic Press, London.

Willcox, J. & Tribe, H.T. (1974). Fungal parasitism in cysts of *Heterodera*. I. Preliminary investigations. *Transactions of the British Mycological Society*, **62**, 585–94.

The potential impact of fungal genetics and molecular biology on biological control, with particular reference to entomopathogens

11.1 Introduction

Many fungi are of potential value for biological control, because of their destructive relationships with host organisms that are major pests (insects, nematodes), fungal plant pathogens, or weeds of agricultural importance. A strain improvement programme requires clearly defined aims, and must take into account the nature of the host–parasite relationship. Investigators have usually focused their attention on traits thought likely to be determinants of pathogenicity, in order ultimately to facilitate the development of strains (by genetic manipulation[1]) that have acquired an enhanced ability to attack the host.

The examples dealt with here are taken from genetic and related studies employing selected entomopathogenic fungi because of the current interest in their application in biocontrol, as well as my direct experience of strain improvement programmes employing *Verticillium lecanii* and *Metarhizium anisopliae* in this laboratory. Only a few such fungi have received detailed genetical attention (Heale, 1982). In addition, although no instances can yet be cited of the application of the rapidly developing techniques of molecular biology in this area, it is considered timely to review how such techniques may be employed and their likely potential.

11.2 Identification of traits determining pathogenicity and efficacy

A detailed understanding of the traits involved in determining pathogenicity[2] is an essential prelude to an effective strain improvement programme for fungi employed in biocontrol, in which a degree of damaging parasitism underlies their mode of action. Many of these traits are probably under multiple gene control, but those determined by single genes and that are capable of genetic manipulation and transfer to otherwise desirable strains are likely to be of considerable interest.

11.2.1. *Spore type*
Genetic studies with fungi are frequently performed with organisms producing uninucleate conidia as typically found in Deuteromycete entomopathogens cultured on solid substrates. However, production of these fungi in liquid fermentation, or in semi-liquid systems, is potentially more efficient. Yeast-like blastospores are often produced in liquid cultures, along with hyphal bodies or fragments, e.g. *Metarhizium anisopliae* (Adamek, 1965), *Beauveria brongniartii (tenella)* (Cartroux *et al.*, 1970), *B. bassiana* (Fargues *et al.*, 1979), *Verticillium lecanii* (Hall & Latgé, 1980) and *Nomuraea rileyi* (Riba & Glandard, 1980). The blastospores are similar to the type produced in the insect haemocoel. They have a shorter shelf-life than conidia, but this may be improved by selection and formulation.

---

[1] 'Genetic manipulation' is used here in the broadest sense, including mutagenesis, parasexual/sexual recombination, somatic hybridisation, transformation, gene cloning and vector transfer.
[2] The general term 'pathogenicity' is employed rather than 'virulence', which implies an understanding of a specific gene-for-gene determined system.

Two commercial products based upon blastospores of *V. lecanii* were developed in the UK by Microbial Resources Ltd (now Novo Industri Ltd): 'Vertalec' (for control of glasshouse aphids) and 'Mycotal' (against glasshouse whitefly). The blastospores are at least as pathogenic as conidia of the same strain[3] to their respective hosts (Hall, 1981; J. Drummond & J.B. Heale, unpublished). Commercial methods often employ conidia for reasons of viability, produced either (1) on a solid substrate, e.g. autoclaved rice grains in cultures of *M. anisopliae* for control of sugar-cane spittlebugs in Brazil, or (2) in a two-stage process, e.g. for *B. bassiana*, comprising first a liquid medium followed by a solid medium, for control of Colorado beetle in the USSR. The detailed methods involved in biomass production of various entomopathogenic fungi have been reviewed by Lisansky & Hall (1983). The value of selection is shown by the report that one particular strain of *Hirsutella thompsonii*, causing epizootics among eriophyid mites, conidiated in submerged culture (van Winkelhoff & McCoy, 1981). Similarly, Paris *et al.* (1985) described two mutant strains of *B. brongiartii* which failed to form blastospores in liquid culture but which still produced conidia under these conditions and retained their pathogenicity. The efficacy of a commercial product is dependent upon a thorough understanding of the way in which the formulated fungal propagules behave in contact with the host and crop, and their full potential will only be realised if the viability of the product is unimpaired and if environmental conditions are favourable for germination and the establishment of infection.

The use of standardised strains purified by single-sporing, and shown to be uninucleate by cytological tests, is essential since isolates initially recovered from the wild may be heterokaryotic. Further, naturally occurring diploid strains of fungi otherwise regarded as regularly haploid have been reported from the wild (Hastie & Heale, 1984). Heterokaryosis and the parasexual cycle in Deuteromycete fungi are frequently proposed as mechanisms involved in the origin of new plant-pathogenic races (Day, 1960; Tinline & MacNeill, 1969; Clarkson & Heale, 1985a, b, c). Eighteen *V. lecanii* isolates from insects, other hosts, or soil, bioassayed against the aphid *Macrosiphoniella sanborni*, produced spores with volumes ranging from 2.2 to 18.3 μm^3, with no direct correlation between size and pathogenicity; all were haploid (Jackson & Heale, 1985; Jackson *et al.*, 1985). Recently, however, we have identified a wild-type diploid strain of this fungus, pathogenic to aphids, with moderately large conidia (13.6 μm^3) (J. Drummond & J.B. Heale, unpublished).

Variation in spore length among populations of *M. anisopliae* is known to be directly correlated with pathogenicity and host-specificity. The large-

[3] 'Strain' is employed to denote a stable, specifically characterised wild type, or one produced by any type of genetic manipulation; 'isolate' is used in every other case.

spored form (13–17 µm in length), termed *M. anisopliae* var. *major* (Tulloch, 1976), is pathogenic to *Oryctes rhinoceros* and other scarabeid beetles, whereas the small-spored (5–8 µm) form, var. *anisopliae*, has been found on over 100 insect hosts. Prior & Arura (1985), however, reported that only small-spored pathogenic isolates were obtained from rhinoceros beetles in Papua New Guinea. Tinline & Noviello (1971) first suggested that var. *major* might be diploid; this is supported by observations in this laboratory involving UV survival, failure to produce auxotrophic markers and acridine staining (K. Samuels & J.B. Heale, unpublished).

11.2.2 *Adhesion and germination*

When the spore contacts the cuticle, adhesion is initiated involving charged groups of both propagule and host surfaces aided by the mucilaginous outer layer of the spore (Fargues, 1984). Specific attachments develop, depending upon initial recognition. Appressoria are commonly observed, although direct penetration by the germ tube also occurs, and both mucus secretion and extracellular enzyme activity can be detected prior to actual germination (Fargues, 1984). The importance of adhesion was shown by Al-Aidroos & Roberts (1978), when they observed that very few conidia of a non-pathogenic mutant of *M. anisopliae* attached to the perispiracular valves of mosquito larvae, compared with high adhesion rates for the pathogenic wild type. Later, Al-Aidroos & Bergeron (1981) reported that a gene determining specific adhesion to the larval surface in the same fungus was linked to that for brown spore colour, and that its segregation could be followed in parasexual crosses. Grula *et al.* (1984) reported that conidia and blastospores (but not germ tubes or hyphae) of *B. bassiana* adhered to chick red blood cells and that the groups responsible for haemaglutinating activity (possibly glycoprotein) were located on the surface of conidia. Such groups could determine the anchorage and orientation of the conidium, facilitating penetration by the emerging germ tube, since spores germinated on an artificial medium showed very little cuticle-penetrating activity after transfer. Thus the ability to adhere to the host surface is an important determinant of pathogenicity which should prove to be amenable to genetic manipulation.

Al-Aidroos & Roberts (1978) selected highly pathogenic strains of *M. anisopliae* by re-isolation from rapidly dying mosquito larvae. These exhibited increased *in situ* germination and invasion. Measurements of *in vitro* germination alone are probably of limited value, however. Thus, although Jackson *et al.* (1985) reported that rapid germination on complete medium was frequently associated with pathogenicity in 18 wild-type isolates of *V. lecanii* tested against the aphid *M. sanborni*, the most pathogenic isolate (LT_{50}/adults: 3 days) had a relatively slow T_{50} germination value on agar of 10 h, compared with 5.5–9.0 h for moderately pathogenic isolates (LT_{50}: 5–9 days). During *in vivo* tests, the most pathogenic isolates consistently displayed the most rapid germination (H. Williams and M. Llewellyn, unpublished). This indicates that they were better adapted to the aphid host surface environment, either by molecular configurations favouring adhesion, response to nutrients (Woods & Grula 1984), and/or enhanced resistance to inhibitors such as fatty acids. Similarly, the use of the vital fluorescent stain 'Uvitex BOPT' (Drummond & Heale, 1985), to observe *in vivo* germination of *V. lecanii*, demonstrated that highly pathogenic isolates germinated more rapidly on whitefly pupae than did relatively non-pathogenic isolates (Drummond *et al.*, 1987). Further, *B. bassiana* mutants with low pathogenicity to *Heliothus zea* either failed to germinate or grew abnormally without penetration (Pekrul & Grula, 1979).

11.2.3 *Penetration and cuticle-degrading enzymes*
The integument consists of an extremely thin epicuticle containing waxy lipids, covering the thick procuticle of chitin embedded in protein (Blomquist, 1984; Jeuniaux, 1984; Locke, 1984). Penetration, as in higher plants attacked by fungi (Maiti & Kolattukudy, 1979), involves *both* enzymes and mechanical pressure (Ferron, 1978; Roberts & Humber, 1981). Secretion of lipase *at the point of penetration* of *Galleria mellonella* and *Bombyx mori* by *Entomophthora coronata* was reported by Gabriel (1968), and of esterase, lipase and N-acetylglucosaminidase from *B. bassiana* on the former host by Michel (1981). Mixtures of proteinases and chitinases caused *in vitro* cuticle dissolution (Smith *et al.*, 1981). Although it is clear that these enzyme systems are involved in penetration by entomopathogenic fungi, the suggestion that any one may act as a *major* determinant of pathogenicity requires compelling evidence, which as yet is lacking.

Of 16 wild-type isolates of *B. brongniartii*, those with high lipase activity on agar were more pathogenic to cockchafer larvae (Paris & Segrétain, 1975), and ten lipase-negative mutants were non-pathogenic (Paris & Ferron, 1979). Further, localised dissolution of the outer waxy layers of the epicuticle of *M. sanborni* by *V. lecanii* has been observed (H. Williams and M. Llewellyn, unpublished). Coating detergent-treated cuticles of *H. zea* with lanolin prevented their *in vitro* degradation by a mixture of proteinases and chitinase produced by *B. bassiana*, unless lipase was present (Smith *et al.*, 1981). However, high lipase activity on agar characterised both pathogenic *and* non-pathogenic, wild-type isolates of *V. lecanii* (Jackson *et al.*, 1985), and two high-lipase *M. anisopliae* strains were of low pathogenicity (Robert & Al-Aidroos, 1985). The high lipase activity on agar of five out of six isolates of *Conidiobolus obscurus*, pathogenic to pea aphid, compared with nil activity of five out of six non-pathogenic isolates, was reversed for several of the isolates when grown in liquid shake cultures

(Latgé *et al.*, 1984). Other technical problems include the secretion of acids by *M. anisopliae* which interfere with assays of both lipases and proteases (Robert & Al-Aidroos, 1985).

All 18 wild-type isolates of *V. lecanii* tested by Jackson & Heale (1985) produced high protease activity on agar irrespective of pathogenicity to the aphid *M. sanborni*, whereas there was a correlation between chitinase levels and pathogenicity. More recently, we have shown that isolates of *M. anisopliae* most pathogenic to *Nilaparvata lugens* are frequently characterised by high endoprotease activities (K. Samuels & J.B. Heale, unpublished). Two chitinase-deficient mutants of *B. brongniartii* were non-pathogenic by surface inoculation but pathogenic by injection of spores into *Melolontha* (Paris *et al.*, 1985). Digestion of *H. zea* cuticles (see above) required sequential activity of proteases and chitinases by *B. bassiana* (Smith *et al.*, 1981), although, perhaps significantly, non-entomopathogenic, contaminant fungi also digested them! *In vitro* cultures of *M. anisopliae*, *B. bassiana* and *V. lecanii* on comminuted fragments of locust cuticle produced a proteolytic complex (endoproteases, aminopeptidase, carboxypeptidase A) within 24 h, and chitinases were detected much later (St. Leger *et al.*, 1986a). A purified endoprotease removed c. 25–30% (w/w) of protein from 'whole' larval cuticle; chitinase alone released 3–4% of the N-acetylglucosamine content of total chitin, and 3.5-fold more after pretreatment with protease (St. Leger *et al.*, 1986b). Regulation of chitinase and chitosanase in *M. anisopliae* was via an inducer/repressor mechanism, the most effective inducers being N-acetylglucosamine and glucosamine respectively; catabolite repression occurred in medium containing carbohydrates, lipids or proteins (St. Leger *et al.*, 1986c). Very recently, St Leger *et al.* (1987a) reported that the major endoprotease activity produced by *M. anisopliae* is a chymoelastase (Pr1) enzyme with serine and histidine residues in the active site. These workers detected Pr1 during cuticle penetration (St. Leger *et al.*, 1987b) and showed that similar chymoelastases are produced by several different entomopathogenic fungi (St. Leger *et al.*, 1987c). In conclusion:

(1) only a limited value can be assigned to any one particular trait among the many that interact and summate to determine pathogenicity;

(2) solid-medium enzyme assays indicate extracellular activity and are subject to catabolite repression and other technical limitations;

(3) enzyme mutants with altered pathogenicity may result from mutations at other sites;

(4) until the molecular composition of the epicuticular lipids and the cuticular proteins masking the chitin of the particular host insect under study is known, it is premature to judge the exact role that cuticle-degrading enzymes may play in host specificity, or even the degree of pathogenicity.

On the living host surface, regulatory mechanisms affecting mechanical pressure and the activity of hyphal-wall-bound as well as 'exported' cuticle-degrading enzymes, operating over microscopic distances, probably determine whether penetration occurs and at what rate. Interestingly, Grula *et al.* (1984) have concluded that the differential ability of *B. bassiana* mutants to penetrate corn earworm larvae integuments could not be correlated with any *in vitro* enzyme activity, and suggested that differences in the attachment and orientation of conidia, determined by specific glycoprotein surface receptors, caused enhanced penetration by emerging germ tubes of the most pathogenic mutants.

11.2.4 *Pathogenicity*
11.2.4.1 *Rate of host death.*
The quantification of pathogenicity is essential for genetical approaches to strain improvement. It is determined by many traits controlled by numerous genes (Jackson *et al.*, 1985); laboratory bioassays under controlled environmental conditions provide an objective method of assessment of its overall expression, although spread in natural populations cannot be predicted reliably. LT_{50} or LT_{95} values, obtained using standard inoculation levels, are employed, with LC_{50} and LC_{95} values over standard times providing additional data.

11.2.4.2 *Toxins and cellulolytic enzymes.*
Host death is not thoroughly understood, but internal colonisation and accompanying nutrient utilisation and digestion are involved in parasitic infections caused by *M. anisopliae*, *B. bassiana* and *V. lecanii*. At least in one case there is strong evidence for toxins, i.e. the cyclodepsipeptide destruxins produced by *M. anisopliae* var. *anisopliae* (Roberts, 1981), detected in infected silkworm larvae at lethal concentrations (Suzuki *et al.*, 1971). They are known to inhibit RNA and DNA synthesis in insect cell cultures (Quiot *et al.*, 1985). Recently, Kaijiang & Roberts (1986) have shown that the large-spored strain, var. *major*, which has a very restricted host range, failed to produce any of the 14 possible destruxins, or produced only traces of destruxin A. Another cyclodepsipeptide, beauvericin, is known from cultures of *B. bassiana*, and toxin production in different isolates has been correlated with pathogenicity to *G. mellonella* (Ferron, 1978), but there is also a contrary report (Champlin & Grula, 1979). An isolate of *V. lecanii* that had lost its pathogenicity on storage failed to produce dipicolinic acid thought to be responsible for insecticidal activity (Claydon & Grove, 1982), but convincing evidence is lacking here.

An important aspect of toxin activity is the extent of localised damage to host cells which may inhibit immune responses. Here, selection and genetic manipulation could enhance toxin production and pathogenicity. Comparatively little has been published on cytolytic enzyme activity by entomopathogenic fungi, but Kucera (1982, 1984) has demonstrated a toxic protease produced by *M. anisopliae* and the counter-role of proteolytic inhibitors in *Galleria mellonella* larvae. Phospholipases with membrane-destroying properties might also be explored from this point of view.

11.2.4.3 *Avoidance of host recognition.* Highly pathogenic wild-type strains are likely to be characterised by surface molecular configurations which fail to be recognised as 'non-self' by the specific insect host. The fact that some entomopathogenic fungi such as *M. anisopliae* attack a very wide range of insects, but that individual strains are more pathogenic to specific hosts, implies a variation in the degree of recognition (as well as the ability to overcome immune responses in different hosts). Such 'recognition signals' should be amenable to genetic manipulation.

Immune responses to fungal ingress include:

(1) localised melanisation during penetration around the invading germ tube, resulting from the presence of phenols and cuticular oxidases;
(2) following penetration, cellular responses involving haemocytic (phagocytic) encapsulation or nodulation, often followed by melanisation enveloping the 'foreign' cells (Lackie, 1979; Dunphy & Nolan, 1982; Boucias & Pendland, 1984; Vey, 1984).

The production of rapidly budding, yeast-like blastospores by many entomopathogenic fungi is probably a two-fold adaptation, involving a degree of altered recognition signals as well as a high rate of spread within the haemocoele, giving the host less time to respond. However, Paris *et al.* (1985) reported that blastospore-negative mutants of *B. brongniartii* were still pathogenic. The production of naked protoplasts during infection by many entomophthorean fungi may again represent an avoidance of recognition (Roberts & Humber, 1984).

In plant and animal hosts attacked by fungi, $\beta1-3$ glucans released from fungal cell walls have been shown to elicit active defence responses (Albersheim & Valent, 1978; Anderson, 1980). Similarly, Soderhall & Unestam (1979) demonstrated that the haemolymph prophenoloxidase system in the crayfish *Astacus astacus* was activated by water-soluble $\beta1-3$ glucans released from fungal cell walls, and Huxham & Lackie (1986) have reported a similar, laminarin-activated system in haemocytes of the male locust *Schistocerca gregaria*.

In recent tests, three isolates of *M. anisopliae* with different levels of pathogenicity to *N. lugens* (brown planthopper of rice) were compared with regard to the immune response of male locust haemocytes (I. Huxham, K. Samuels, A. Lackie & J.B. Heale, unpublished). Using haemocytes prepared as a monolayer in a petri plate, two procedures were followed: (a) haemocytes were treated *in vitro* with a water-soluble extract of 24-h germinating spores; (b) locusts were injected *in vivo* with a hyphal wall preparation 6 h prior to removal of haemocytes for monolayering. In both cases, in the presence of laminarin, the most pathogenic isolate (against *N. lugens* by inoculation and *S. gregaria* by injection) induced the least activation of the dopa-polyphenoloxidase system leading to melanisation which would otherwise inhibit invasion by the pathogen.

11.2.5 *Sporulation and spread*

Rapid death of the host, accompanied by efficient utilisation of substrates, and the early production of large numbers of 'readily-released' spores at the surface of the cadaver, will undoubtedly facilitate spread of the pathogen through the insect population if environmental factors are not limiting. In this laboratory, we have attempted to select strains of both *V. lecanii* and *M. anisopliae* with improved sporulation traits.

V. lecanii is wet-spored, with conidia produced in mucilaginous heads which are dispersed by a water-splash mechanism or contact with a moving insect. A mutant strain selected after UV irradiation released 4.4×10^5 spores per leaf disc (after shaking three tomato leaf discs in 2 ml of sterile distilled water for 20 s, each disc supporting 10 infected whitefly (*Trialeurodes vaporariorum*) pupae, 7 days after inoculation). This represents a 1×10^5 increase in spore release over the parental isolate (J. Drummond & J.B. Heale, unpublished), and was still expressed after 20 passages on agar. Although showing a reduction in pathogenicity, presumably due to mutation at loci determining other traits, this strain could be employed as a stable, enhanced sporulation genotype in attempts at genetic manipulation for strain improvement.

M. anisopliae produces dry spores with a strongly hydrophobic nature, which limits the rate of epizootic spread in insect populations. Diauxotrophic mutants have been selected with an enhanced ability for: (a) spore production, and (b) the ease with which they are detached from their aggregates or hyphal contacts by gentle swirling in sterile distilled water containing 0.05% Tween 80 (surfactant) as opposed to the more violent agitation commonly used in preparing conidial suspensions of dry-spored fungi. Two enhanced sporulation 'producers' and 'releasers', derived from different *N. lugens* wild-type isolates from the Philippines, have been employed recently in a parasexual cross and selected recombinants have been tested for their pathogenicity in laboratory bioassays here and in one case for epizootic spread under field conditions at the International Rice Research Institute (IRRI) in the Philippines (K. Samuels, M. Rombach & J.B. Heale, unpublished).

Rates of spread determined under laboratory conditions are unlikely to predict reliably the outcome of a field or glasshouse epizootic caused by any one strain, although they may be useful as a general indication. The introduction of a cadaver infected by *M. anisopliae* to a healthy population of *N. lugens* maintained under standard conditions, followed by estimates of population mortality rate, may be employed for this purpose, rather than direct inoculation as used in standard bioassays.

11.2.6 *Stress tolerance*

The success of biocontrol using entomopathogenic fungi in temperate regions is limited strictly by adverse low humidity and temperature conditions. During glasshouse control of pests such as aphids and whitefly, higher humidity (>95% r.h.) periods frequently alternate with dry intervals (<75% r.h.), especially at midday, which can affect adversely the development of an epizootic. *V. lecanii* requires >93% absolute water activity for spore germination and growth (A. Gillespie, personal communication). A recent report by Milner & Lutton (1986) confirmed that little spread of this fungus occurred at or below 93% r.h. in populations of the aphid *Myzus persicae* sprayed initially with a commercial preparation of 'Vertalec'. Recently, we have selected a wild-type strain of *V. lecanii* that was more pathogenic to glasshouse whitefly (*T. vaporariorum*) than the commercial strain (employed in 'Mycotal') after an initial 16 h at >95% r.h. (simulating a late-afternoon application of the fungus) followed by extended periods of low humidity (70% r.h.). Examination of infected pupae indicated that this strain had developed more rapidly during the first 16 h at high humidity than the other isolates tested (Drummond *et al.*, 1987), suggesting a degree of 'escape' from the following low humidity stress, but more recent tests suggest an additional ability of this strain to continue to grow at low r.h.

Hall (1981) determined optimal germination for a large number of *V. lecanii* isolates at between 20 and 25°C, with very little germination occurring at 11°C, but above 30°C different isolates showed considerable variation in response, one growing even at 36°C, indicating the scope for selection here. A low-temperature strain of *M. anisopliae* from New Zealand has a growth optimum of 22°C and fails to grow at 28°C, the optimum for tropical/sub-tropical isolates tested (K. Samuels & J.B. Heale, unpublished). A strategy which could be employed here would be to develop strains with overlapping temperature optima to be applied as 'mixtures'.

11.3 **Genetic manipulation**

11.3.1 *Selection*

It has been suggested that selection of wild-type isolates recovered from other insects, or from other populations of the target pest in widely different geographical areas, is more likely to be effective than choosing an isolate which has reached an equilibrium with the pest population (Roberts & Humber 1981). Such diverse isolates may represent divergent races which have evolved unique genotypes and at the same time may have developed incompatibility systems which prevent gene flow occurring normally, even in coincident populations. Such barriers can usually be overcome using genetic manipulation by 'forcing' heterokaryons, somatic hybridisation employing protoplasts, or various transformation procedures. This makes it possible to recombine traits between (1) highly pathogenic isolates with obvious epizootic capability towards the target pest, by direct re-isolation from 'field' populations, and (2) a diverse range of other isolates or strains chosen for specific characteristics, rather than proven pathogenicity or epizootic capability towards the target insect. Currently, we are attempting to do this by 'crosses' between highly pathogenic isolates of *M. anisopliae* obtained from *N. lugens* in the Philippines and isolates of the same pathogen from sugar cane spittlebug in Brazil that have enhanced sporulation characteristics and have been used commercially for these pests.

There are conflicting reports of loss of pathogenicity in various entomo-pathogenic fungi with increasing serial passaging on artificial media, particularly in the case of *M. anisopliae* (Tinline & Noviello, 1971; Daoust & Roberts, 1982). *V. lecanii*, on the other hand, appears to be more stable under these conditions (Hall, 1981; J.B. Heale, unpublished data). Nevertheless, it is good practice to prepare numerous liquid nitrogen subcultures of a selected strain (after single-sporing) to avoid variation. Similarly, all isolates received from other laboratories require immediate single-sporing, bioassay, fresh re-isolation and long-term storage.

Cycles of selection within infected populations in the laboratory by re-isolation from 'early dying' insects can lead to enhancement of pathogenicity, as shown by tests with moderately pathogenic aphid isolates of *V. lecanii* against *M. sanborni*; however, such procedures did not improve the level of pathogenicity above that of the most pathogenic wild-type isolate (C. Jackson & J.B. Heale, unpublished). Al-Aidroos & Roberts (1978) were able to select strains with enhanced pathogenicity by re-isolation of *M. anisopliae* from inoculated mosquito larvae which died within 2 days; these strains exhibited more rapid germination and invasion of the host, increased sporulation and destruxin production. Later, Daoust & Roberts (1982) observed that much of the natural variation in pathogenicity of 47 isolates of *M. anisopliae* (from other insect hosts from diverse regions) to mosquito larvae was due to different spore viabilities, but that passage through the latter host increased pathogenicity. In one case, passage through an alternate host, namely brown planthopper (*N. lugens*), raised its pathogenicity to mosquito larvae significantly. These authors suggested that parasitic selection pressures probably led to an enhanced ability of conidia to germinate rapidly, thus increasing their invasive capacity.

11.3.2 *Mutagenesis*
There are three major reasons for employing mutants in genetical studies with Deuteromycete fungi:

(1) production of specific nuclear 'markers' which allow the detection of recombinational events, segregation patterns and linkage groups;
(2) auxotrophs can be utilised to 'force' heterokaryons on minimal medium and to select for heterozygous diploids and recombinant prototrophic haploids obtained in crosses between complementary diauxotrophic parent strains;
(3) enhancement or lowering of activity affecting specific traits, e.g. germination rate, sporulation density, cuticular-degrading enzymes, etc. allows investigation of possible determinants of pathogenicity.

Mutagenesis, followed by screening on the target population, re-isolation from early dying insects and further tests for pathogenicity are alternative strategies for strain improvement.

However, mutagenesis of a particular strain may cause changes at loci other than that detected initially, and the use of auxotrophs is complicated by pleiotropic effects. For example, auxotrophic strains are usually less pathogenic than their parent wild type (Jackson & Heale, 1986). These effects are reversed frequently in prototrophic, recombinant haploids derived from crosses. Alternative markers include drug resistance and temperature-sensitive mutants. The types of mutagen used in various studies involving entomopathogenic fungi are shown in Table 11.1. Total isolation methods have been most frequently used for auxotrophs but, to improve efficiency and percentage yield, early selection against normally growing wild types in minimal medium has been successfully employed, either by filtration or by the use of the antifungal antibiotic, nystatin (Table 11.1).

In this laboratory, we have found NTG (N-methyl-N^1-nitro-N-nitroso-guanidine) to be a more effective mutagen for *M. anisopliae* and *V. lecanii*, producing auxotrophs with greater stability than UV. NTG is an alkylating agent which induces a high frequency of mutations at or close to DNA replication points by mispairing as a result of tautomerisation of the alkylated bases. Although it is a strong mutagen, some strains are resistant to its effects (Cero-Olmedo & Ruiz-Vazquez, 1979; Jackson, 1984).

Auxotrophs produced most commonly in these studies include *ade, lys, phe, arg, leu, met, val, tryp, thi, cys, pyr, bi, nic, pab* and *pan* (symbols according to Demerec *et al.*, 1966). Spore colour mutants have been used extensively in parasexual crosses involving *M. anisopliae*. Here wild-type conidia are green as in *Aspergillus nidulans* and a range of EMS (ethyl methane sulphonate)-induced mutants have been described by Al-Aidroos & co-workers (Table 11.1), including 'brown', 'tan' and 'yellow', as well as 'lemon' induced by UV. Messias & Azevedo (1980) and Riba *et al.* (1985) have employed UV-induced 'yellow', 'white' and 'pale vinaceous' (violet), spore colour markers. The immediate advantage of these mutants in parasexual crosses is that when complementation occurs in heterozygous diploid

Table 11.1. Mutagenic studies in entomopathogenic fungi

Mutagen[a]	Fungus	Method	Mutants	Author(s)
UV	M. anisopliae	Filtration enrichment Filtration enrichment Filtration enrichment	Auxotrophs Spore colour Mal. green[b]	Tinline & Noviello (1971)
UV	M. anisopliae	Total isolation Total isolation	Auxotrophs Spore colour	Messias & Azevedo (1980)
UV	M. anisopliae	Filtration enrichment Filtration enrichment	Auxotrophs Spore colour	Silveira & Azevedo, (1984); Riba et al. (1985)
UV + NTG	M. anisopliae	Total isolation Total isolation Total isolation Total isolation Total isolation	Auxotrophs Spore colour Spore density Spore release Ex. enzymes[c]	Samuels & Heale, unpublished
EMS EMS	M. anisopliae M. anisopliae	Total isolation Total isolation Total isolation	Spore density Auxotrophs Spore colour	Al-Aidroos & Roberts (1978) Al-Aidroos (1980); Al-Aidroos & Bergeron (1981); Bergeron & Al-Aidroos (1982); Magoon & Al-Aidroos (1984, 1986)
UV + NTG	B. brongniartii	Nystatin selection Nystatin selection Nystatin selection Nystatin selection	Auxotrophs Blastosp.[d] chitinase pigment	Paris & Ferron (1979); Paris et al. (1985)
UV + NTG	V. lecanii	Total isolation Total isolation	Auxotrophs Spore density	Jackson & Heale (1986); Drummond & Heale, unpublished
		Total isolation Total isolation	Spore release Ex. enzymes[b]	

[a] UV, NTG, EMS: see text.

[b] Mal. green: malachite green resistance.

[c] Ex. enzymes: chitinase, protease, lipase enhanced and deficient producers.

[d] Blastosp: blastospore negative.

colonies derived from appropriate pairs of colour mutant parents (as well as in recombinant haploids carrying the two corresponding wild-type alleles), the wild-type green colour is expressed and can be selected for visually. Recently, Magoon & Al-Aidroos (1986) have suggested that there are two metabolic pathways for spore pigmentation here, with 'lemon', 'tan' and 'yellow' in one, and 'brown' in a separate pathway, based upon epistatic relationships in heterozygous diploids and recombinant haploids. 'Red' haploid recombinants were recovered from diploids obtained from crosses of 'tan' with 'brown' strains, and from other crosses; 'mustard' and 'orange' phenotypes were expressed in haploid recombinants.

Spontaneous acriflavine resistance was selected for in *M. anisopliae* by Al-Aidroos (1980); later studies showed that independent *acr*-resistant mutant alleles were either dominant or expressed incomplete dominance in heterozygous diploids (Bergeron & Al-Aidroos, 1982; Magoon & Al-Aidroos, 1986).

11.3.3 *Parasexual recombination*

Parasexual mechanisms of recombination not only enable genetic analysis in deuteromycete fungi, but also provide an important tool in strain improvement for those organisms with industrial or other applications, such as entomopathogens (Heale, 1982). The parasexual cycle (Pontecorvo, 1959) consists essentially of the following steps:

(1) Haploid homokaryons of complementary genotype anastomose, followed by nuclear migration, thus establishing a heterokaryon.
(2) Rare nuclear fusions may occur, resulting in heterozygous diploid nuclei.
(3) Heterozygous diploid ($2n$) nuclei may undergo regular mitoses, but they also produce irregular genotypes — either by mitotic recombination or by mitotic non-disjunction, leading via successive aneuploid states ($2n$-1, $2n$-2, etc. . . .) to a haploid number of chromosomes (n). The latter process (haploidisation) results in a random 'shuffling' of the chromosomes of the two original parents, but markers on the same chromosome show complete linkage unless separated by a mitotic cross-over event.

Numerous parasexual 'crosses' involving various combinations of auxotrophs and spore colour markers have been reported for *M. anisopliae* by Azevedo & co-workers in Brazil, and by Al-Aidroos and co-workers in Canada (see Table 11.1 citations for references). In these studies, various haploidising agents, including chloroneb, benlate and botran, have been employed to treat the heterozygous diploids first formed, but in this laboratory we have found such diploids to be very unstable and do not require such agents.

Linkage studies and chromosome mapping via parasexual analyses are very imprecise, owing to the possible coincidence of mitotic crossing-over in progeny that have undergone haploidisation, and particularly because of undetected differences in rates of mitoses and selective advantages in hyphal nuclei of varying genotype prior to conidial production and sampling on to selective media. However, it is possible to assign probable linkage groups by detecting a high frequency of linkage between selected pairs of markers in segregants produced by haploidisation from heterozygous diploids. Thus, Al-Aidroos & Bergeron (1981) reported linkage between a gene for spore adhesiveness and yellow spore colour in *M. anisopliae*, and recently Magoon & Al-Aidroos (1986) have assigned ten markers to five linkage groups (corresponding to five chromosomes in the haploid state) in the same fungus.

Concerning the inheritance of pathogenicity via parasexual recombination, Riba *et al.* (1985) found that a heterozygous diploid of *M. anisopliae*, produced from a 'cross' between complementary colour and auxotrophic strains originating from an identical parent isolate, was as pathogenic to mosquito larvae as the parent, indicating full complementation for numerous pathogenicity genes. However, haploid prototrophic recombinants showed reduced pathogenicity, as did the parental auxotrophs. Later, Messias *et al.* (1986) found that a diploid from a different cross involving strain E9 was as pathogenic to *Rhodnius prolixus* (a vector of the fatal 'Chagas' disease in South America) as the parental isolate. Again using *M. anisopliae*, and by selection cycles for enhanced sporulation traits employing mutagenesis and recombination, we have obtained a prototrophic haploid recombinant from a cross between colour and diauxotrophic parent strains derived from two independent wild types, both originating from rice brown planthopper in the Philippines; the recombinant was intermediate in pathogenicity to the wild-type parents in the laboratory but it produced 1000 times the number of 'readily releasable' spores as the 'best' of the two parents: this strain is being tested for its epizootic efficacy towards *N. lugens* at IRRI in the Philippines (K. Samuels, M. Rombach & J.B. Heale, unpublished).

We have also investigated 70 different crosses, employing 21 diauxotrophs derived from seven wild-type isolates of *V. lecanii* obtained from the target aphid (*M. sanborni*), other insects, and non-insect hosts and soil from seven different countries (Jackson & Heale, 1986). Complete incompatibility was expressed between certain isolates, e.g. an isolate from a fern in The Netherlands and the other from an alternative host insect in Czechoslovakia. Other pairs were highly compatible, e.g. a *Myzus persicae* (aphid) isolate from India crossed with a highly pathogenic *M. sanborni* (aphid) isolate from the UK. Most highly incompatible paired isolates yielded at least a few heterozygous diploid conidia. Evidence for diploidy was provided by DNA Feulgen microdensitometry, prototrophy, and segregation resulting in recombinant haploids. Spore volumes obtained by Coulter counter analysis were not a reliable indicator if used for crosses between parents with widely different spore sizes. We have also employed similar approaches in testing parasexual recombinants of this fungus against whitefly, but additionally heterozygous diploids were screened directly on the target insect followed by early re-isolation. Generally, heterozygous diploids were near to the parent wild-type isolates in their pathogenicity, and were enhanced relative to the auxotrophic strains employed, the latter expressing either pleiotropic reductions or undetected mutagenic changes. Recombinant haploids showed intermediate or reduced pathogenicity compared with parental isolates. This could result from disruption of gene clusters controlling pathogenicity by mitotic recombination, as suggested by Clarkson & Heale (1985b).

11.3.4 *Somatic hybridisation*

Incompatibility expressed at the hyphal wall, thus preventing hyphal fusions between different strains, can be circumvented by the use of protoplast fusion (Peberdy, 1979). Techniques for fungal protoplast isolation and fusion have been developed (Anné & Peberdy, 1975; Ferenczy, 1981), and we have produced successful protocols, both for *V. lecanii* and *M. anisopliae* (Jackson & Heale, 1987; K. Samuels & J.B. Heale, unpublished). Heterozygous diploids (and haploid recombinants) of the former were obtained employing protoplast fusion between complementary diauxotrophic strains, where previous attempts using hyphal anastomosis had failed. In a number of cases, however, neither method resulted in diploids, indicating that incompatibility here was determined by nuclear or cytoplasmic systems as described by Typas (1983). Nevertheless, protoplast fusion has considerable potential for genetic recombination between different species, and gene transfer has been reported for several interspecific hybrids in the fungi including *P. chrysogenum* × *P. roqueforti* (Anné & Peberdy, 1976), *A. nidulans* × *A. fumigatus* (Ferenczy, 1976), and *V. albo-atrum* × *V. dahliae* (Typas, 1983).

11.3.5 *'Dead donor' technique*

As a further development of protoplast biotechnology for strain improvement, we are developing the so-called 'dead donor' technique. The principal reason is that one of the parental lines used in the 'cross' does not required mutagenesis, thus avoiding unwanted mutations occurring in a highly selected, wild-type strain. The method can involve a hybrid (heterologous) fusion between heat-inactivated, wild-type protoplasts of one strain and those of an untreated, auxotrophic mutant strain with a trait likely to improve epizootic efficacy, e.g. enhanced sporulation. Fusion products are screened on minimal medium on which neither (a) inactivated, wild-type nor (b) auxotrophic, unfused, parental protoplasts will regenerate. Selection operates for heterologous fusions, producing heterokaryons and heterozygous diploids. This results in haploid recombinants by mitotic recombination/haploidisation, provided that inactivation does not inhibit any such steps. One problem to resolve is whether prototrophs on minimal medium are actually due to recombination or to transfer of the undamaged nucleus of the heat-treated, wild-type protoplast into the cytoplasm of the untreated, auxotrophic, recipient protoplast. Further refinements of the technique involve alternative inactivation procedures (UV, irreversible chemical inhibitors), and employment of drug resistance markers, e.g. acriflavine, benoyml (Bradshaw & Peberdy, 1984).

11.3.6 *Vectors for fungal transformation*

It is now possible to manipulate specific genes encoding particular traits by the use of special plasmid vectors and the application of the rapidly developing techniques of molecular biology. For entomopathogenic fungi it is still not clear which individual genes should be targeted initially. However, the restoration of pathogenicity in non-pathogenic mutants, derived initially from pathogenic wild-type strains by transformation vectors carrying specific DNA fragments derived from the wild-type genome, offers an objective approach to a definitive understanding of pathogenicity, in the field of both insect and plant pathology. Likely gene candidates among entomopathogens include those determining spore coat glycoproteins/recognition responses/adhesion, regulation of lipase and protease activity, and toxin production.

The first fungus in which DNA-mediated transformation was reported was the yeast *Saccharomyces cerevisiae* (Hinnen *et al.*, 1978) and, for a filamentous fungus, *Neurospora crassa* (Case *et al.*, 1979). More recently, a number of successful transformation systems have been described for *Aspergillus nidulans* (Hynes, 1986). An ideal approach to the molecular dissection of pathogenicity in a plant-pathogenic fungus has been presented by Yoder *et al.* (1986). These workers chose the maize pathogen *Cochliobolus heterostrophus* (anamorph: *Bipolaris maydis* = *Helminthosporium maydis* = *Dreschlera maydis*), which is inhibited by the antibiotic hygromycin B. Wild-type sensitive protoplasts were transformed using a plasmid vector constructed from promotor sequences from *Cochliobolus* and a *hygB* gene from *E. coli* which encodes a hygromycin B-phosphotransferase inactivating the drug.

Similarly, the amds$^+$ gene from *A. nidulans*, which encodes the enzyme acetamidase that permits growth on acetamide as the sole nitrogen source, has been cloned into plasmid pBR322 by Hynes *et al.* (1983) to produce the recombinant plasmid p3SR2. Yoder and colleagues (1986) have transformed protoplasts of *Cochliobolus* using this plasmid, screening on an acetamide medium. The next step will be to construct cosmid vectors using either *amdS* or *hygB* as selectable markers and containing large (35–45 kb) DNA sequences, so that only about 2000 transformants will be required to screen the entire *Cochliobolus* wild-type genomic library for the presence of a particular virulence gene. For this an avirulent strain will be transformed with a cosmid vector constructed as above, the selected transformants being tested on the the host plant to detect restoration of virulence. This will pinpoint the specific DNA sequence carrying the virulence gene, which can then be sequenced and its protein product synthesised and, it is hoped, identified. The same research group (Yoder *et al.*, 1986) has also reported the successful transfer, and its expression in *A. nidulans*, of a cloned virulence gene (*PDA*) from *Nectria haematococca* (attacking pea), which encodes the enzyme cytochrome P-450 monooxygenase that demethylates the phytoalexin pisatin.

Similar approaches can be employed with entomopathogenic fungi. Further, it would be possible to transform mutants lacking specific traits such as lipase or protease activity with appropriate vectors containing selectable markers and promotor signals, plus fragments of the relevant wild-type genomic library, followed by screening of transformants for restoration of enzyme activity, as well as bioassays on the insect host to assess effects on pathogenicity. This would be particularly appropriate for strains with near optimum pathogenicity but shown to be deficient in just one transformable trait.

11.3.7 Standardisation
It is essential to identify the effects of specific improved strains when released, particularly in areas where wild types of the same species are likely to cause some degree of infection in the insect populations concerned. Protection of patents involving genetically manipulated strains may also be an important consideration. Additionally, strain standardisation is a vital aspect of purity during production. Individual strains can be characterised by electrophoretic separation of phosphatases and esterases (De Conti *et al.*, 1980; Da Silva & Azevedo, 1983), or gas chromatography (Messias *et al.*, 1983).

Further likely developments include the typing of strains by incorporation of specific plasmids whose presence can readily be detected as well as the applications of DNA restriction fragment enzyme polymorphism which are already proving invaluable in forensic analysis (Gill *et al.*, 1985).

11.3.8 Release of 'genetically manipulated' strains
The controversial and complex matter of legislation regarding the release of genetically manipulated microorganisms into the natural environment is currently being actively debated. The topic is beyond the scope of this re-

view, but one may conclude that the majority of the genetic manipulation techniques described above, including mutagenesis, and the parasexual and somatic hybridisation/'dead donor' methods involving recombination of genes between different strains *of the same species* (rather than single gene cloning, and transfer by vector to an *unrelated genome*), will prove 'environmentally' acceptable; they will therefore probably avoid legislation. Speculatively, it remains to be seen to what extent safety testing and legislation would delay or prevent the release of any hypothetical novel genotypes incorporating the results of single gene transfer between unrelated organisms. e.g. if a toxin gene produced by the bacterium *B. thuringiensis* were to be successfully transferred and expressed by an entomopathogenic fungus.

Acknowledgements

Thanks are due to Dr. S. Lisansky and Dr R. Quinlan for helpful discussion. I also acknolwedge the financial support of the SERC and the collaboration of Dr A. Gillespie, Institute of Horticultural Research, Littlehampton.

References

Adamek, L. (1965). Submerged cultivation of the fungus *Metarhizium anisopliae*. *Folia Microbiologia (Praha)*, **10**, 255–7.

Al-Aidroos, K. (1980). Demonstration of a parasexual cycle in the entomopathogenic fungus *Metarhizium anisopliae*. *Canadian Journal of Genetics and Cytology*, **22**, 309–14.

Al-Aidroos, K. & Bergeron, D. (1981). Use of the parasexual cycle to relate spore adhesiveness to virulence in the entomopathogenic fungus *Metarhizium anisopliae*. *Genetics Society of America*, **6**, 52 (Abst.).

Al-Aidroos, K. & Roberts, D.W. (1978). Mutants of *Metarhizium anisopliae* with increased virulence toward mosquito larvae. *Canadian Journal of Genetics and Cytology*, **20**, 211–19.

Albersheim, P. & Valent B. (1978). Host pathogen interactions in plants. *Journal of Cell Biology*, **78**, 627–43.

Anderson, A.J. (1980). Studies on the structure and elicitor activity of fungal glucans. *Canadian Journal of Botany*, **58**, 2343–8.

Anné, J. & Peberdy, J.F. (1975). Conditions for induced fusions of fungal protoplasts in polyethylene solutions. *Archives of Microbiology*, **105**, 201–5.

Anné, J. & Peberdy, J.F. (1976). Induced fusion of fungal protoplasts following treatment with polyethylene glycol. *Journal of General Microbiology*, **92**, 413–17.

Bergeron, D. & Al-Aidroos, K. (1982). Haploidization analysis of heterozygous diploids of the entomogenous fungus *Metarhizium anisopliae*. *Canadian Journal of Genetics and Cytology*, **24**, 643–51.

Blomquist, G.J. (1984). Cuticular lipids of insects. In: *Infection Processes of Fungi*, a Bellagio Conference, 1983, eds D. W. Roberts & J. R. Aist, pp. 54–60. The Rockefeller Foundation, New York.

Boucias, D. G. & Pendland, J.C. (1984). Host recognition and specificity of entomo-
    pathogenic fungi. In: *Infection Processes in Fungi*, a Bellagio Conference, 1983,
    eds D. W. Roberts & J. R. Aist, pp. 185–96. The Rockefeller Foundation, New
    York.
Bradshaw, R. E. & Peberdy, J. F. (1984). Protoplast fusion in *Aspergillus:* selec-
    tion of interspecific heterokaryons using fungal inhibitors. *Journal of Microbio-
    logical Methods*, **3**, 27–32.
Cartroux, G., Calvez, J., Ferron, P. & Blachere, H. (1970). Mise au point d'une
    préparation entomopathogène a bàse de blastospores de *Beauveria tenella*
    (Delaw) Siemasko pour la lutte microbiologique contre le ver blanc. *Annales de
    Zoologie — Ecologie Animale*, **2**, 281–94.
Case, M.E., Schweizer, M., Kushner, S.R. & Giles, N.H. (1979). Efficient trans-
    formation of *Neurospora crassa* by utilizing hybrid plasmid DNA. *Proceedings of
    the National Academy of Science, USA*, **76**, 5259–63.
Cero-Olmedo, E. & Ruiz-Vazquez, R. (1979). Nitrosoguanidine mutagenesis. In:
    Genetics of Industrial Microorganisms, ed. K.D. Macdonald, pp. 213–27.
    Academic Press, London.
Champlin, F.R. & Grula, E.A. (1979). Non-involvement of beauvericin in the
    entomopathogenicity of *Beauveria bassiana*. *Applied and Environmental Micro-
    biology*, **37**, 1122–5.
Clarkson, J.M. & Heale, J.B. (1985a). Virulence and colonisation studies on wild
    type and auxotrophic isolates of *Verticillium albo-atrum* from hop. *Plant Path-
    ology*, **34**, 119–28.
Clarkson, J.M. & Heale, J.B. (1985b). Heterokaryon compatibility and genetic
    recombination within a host plant between hop wilt isolates of *Verticillium albo-
    atrum*. *Plant Pathology*, **34**, 129–38.
Clarkson, J.M. & Heale, J.B. (1985c). A preliminary investigation of the genetics
    of pathogenicity in hop wilt isolates of *Verticillium albo-atrum*. *Transactions of
    the British Mycological Society*, **85**, 345–50.
Claydon, N. & Grove, J.F. (1982). Insecticidal secondary metabolic products from
    the entomogenous fungus *Verticillium lecanii*. *Journal of Invertebrate Pathology*,
    40, 413–18.
Daoust, R.A. & Roberts, D.W. (1982). Virulence of natural and insect-passaged
    strains of *Metarhizium anisopliae* to mosquito larvae. *Journal of Invertebrate
    Pathology*, **40**, 107–17.
Da Silva, P. & Azevedo, J.L. (1983) Esterase pattern of a morphological mutant of
    the deuteromycete *Metarhizium anisopliae*. *Transactions of the British Mycologi-
    cal Society*, **81**, 161–3.
Day, P. (1960). Variation in phytopathogenic fungi. *Annual Review of Microbio-
    logy*, **14**, 1–16.
De Conti, E., Messias, C.L., De Souza, H.M.L. & Azevedo, J.L. (1980). Electro-
    phoretic variation in esterases and phosphatases in 11 wild type strains of *Metar-
    hizium anisopliae*. *Experientia*, **36**, 293–4.
Demerec, M., Adelberg, E.A., Clark, A.J. & Hartman, P.E. (1966). A proposal
    for a uniform nomenclature in bacterial genetics. *Genetics*, **54**, 61–76.
Drummond, J. & Heale, J.B. (1985). Vital staining of the entomopathogen *Verti-
    cillium lecanii* on a live insect host. *Transactions of the British Mycological
    Society*, **85**, 171–3.
Drummond, J., Heale, J.B. & Gillespie, A.T. (1987). Germination and the effect
    of reduced humidity on the expression of pathogenicity in *Verticillium lecanii*
    against the glasshouse whitefly *Trialeurodes vaporariorum*. *Annals of Applied
    Biology*, **111**, 193–201.

Dunphy G.R. & Nolan, R.A. (1982). Cellular immune responses of spruce bud-
worm larva to *Entomophthora egressa* protoplasts and other test particles.
Journal of Invertebrate Pathology, **39**, 81–92.

Fargues, J. (1984). Adhesion of the fungal spore to the insect cuticle in relation to
pathogenicity. In: *Infection Processes of Fungi*, a Bellagio Conference, 1983, eds
D.W. Roberts & J.R. Aist, pp. 90–110. The Rockefeller Foundation, New
York.

Fargues, J., Robert, P.H. & Reisinger, O. (1979). Formulation des production de
masse de l'hypomycete entomopathogène *Beauveria* en vue des applications phy-
tosanitaires. *Annales de Zoologie–Ecologie Animale*, **11**, 247–57.

Ferenczy, L. (1976). Some characteristics of intra- and inter-specific fusion pro-
ducts of *Aspergillus nidulans* and *Aspergillus fumigatus*. In: *Cell Genetics in
Higher Plants*, eds D. Dudits, G.L. Fakas & P. Maliga, pp. 55–62. Pergamon,
Oxford.

Ferenczy, L. (1981). Microbial protoplast fusion. In: *Genetics as a Tool in Microbiology*, Society for General Microbiology, Symposium 31, eds S.W. Glover &
D.A. Hopwood, pp. 1–34.

Ferron, P. (1978). Biological control of insect pests by entomogenous fungi. *Annual
Review of Entomology*, **23**, 409–42.

Gabriel, B.P. (1968). Histochemical study of the insect cuticle infected by the
fungus *Entomophthora coronata*. *Journal of Invertebrate Pathology*, **11**, 82–9.

Gill, P., Jeffreys, A.J. & Werrett, D.J. (1985). Forensic application of DNA 'finger-
prints'. *Nature, London*, **318**, 577–9.

Grula, E.A., Woods, S.P. & Russell, H. (1984). Studies utilizing *Beauveria bassiana*
as an entomopathogen. In: *Infection Processes of Fungi*, a Bellagio Conference,
1983, eds D.W. Roberts & J.R. Aist, pp. 147–52. The Rockefeller Foundation,
New York.

Hall, R.A. (1981). *Verticillium lecanii* as a microbial insecticide of aphids and
scales. In: *Microbial Control of Pests and Plant Diseases 1970–1980*, ed. H.D.
Burges, pp. 483–98. Academic Press, New York.

Hall, R.A. & Latgé, J.P. (1980). Étude de quelques facteurs stimulant la formation
in vitro des blastospores de *Verticillium lecanii* (Zimm) Viegas. *Comptes Rendus
hebdomadaires des Séances de l'Academie de Sciences*, **291**, 75–8.

Hastie, A.C. & Heale, J.B. (1984). The genetics of *Verticillium*. *Phytopathologia
Mediterranea*, **XXIII**, 130–62.

Heale, J.B. (1982). Genetic studies on fungi attacking insects. In: *Invertebrate
Pathology and Microbial Control*. Proceedings IIIrd International Society for
Invertebrate Pathology Meeting, Brighton, Sussex, pp. 25–7.

Hinnen, A., Hicks, J.B. & Fink, G.R. (1978). Transformation of yeast. *Proceedings of the National Academy of Science, USA*, **75**, 1929–33.

Huxham, I.M. & Lackie, A.M. (1986). A simple visual method for assessing the
activation and inhibition of phenoloxidase production by insect haemocytes *in
vitro*. *Journal of Immunological Methods*, **94**, 271–7.

Hynes, M.J. (1986). Transformation of filamentous fungi. *Experimental Mycology*,
10, 1–8.

Hynes, M.J., Corrick, C.M. & King, J.A. (1983). Isolation of genomic clones containing the *amdS* gene of *Aspergillus nidulans* and their use in the analysis of
structural and regulatory mutations. *Molecular and Cellular Biology*, **3**, 1430–9.

Jackson, C.W. (1984). Genetical studies on the entomopathogenic fungus *Verticillium lecanii* (Zimm.) Viegas. PhD Thesis, University of London.

Jackson, C. W. & Heale (1985). Relationship between DNA content and spore
volume in sixteen isolates of *Verticillium lecanii* and two new diploids of *V. dahliae*

(= *V. dahliae* var. *longisporum* Stark). *Journal of General Microbiology*, **131**, 3229–36.

Jackson, C.W. & Heale, J.B. (1987). Parasexual crosses by hyphal anastomosis and protoplast fusion in the entomopathogen *Verticillium lecanii* (Zimm.) Viegas. *Journal of General Microbiology*, **133**, 3537–47.

Jackson, C.W., Heale, J.B. & Hall, R.A. (1985). Traits associated with virulence to the aphid *Macrosiphoniella sanborni* in eighteen isolates of *Verticillium lecanii*. *Annals of Applied Biology*, **106**, 39–48.

Jeuniaux, C. (1984). Some aspects of the biochemistry of the chitin protein complex in the insect procuticle. In: *Infection Processes of Fungi*, A Bellagio Conference, 1983, eds D.W. Roberts & J.R. Aist, pp. 61–73. The Rockefeller Foundation, New York.

Kaijiang, L. & Roberts, D.W. (1986). The production of destruxins by the entomogenous fungus *Metarhizium anisopliae* var. *major*. *Journal of Invertebrate Pathology*, **47**, 120–2.

Kucera, M. (1982). Protease inhibitor of *Galleria mellonella* acting on the toxic protease from *Metarhizium anisopliae*. *Journal of Invertebrate Pathology*, **38**, 33–8.

Kucera, M. (1984). Partial purification and properties of *Galleria mellonella* larvae proteolytic inhibitors acting on *Metarhizium anisopliae* toxic protease. *Journal of Invertebrate Pathology*, **43**, 190–6.

Lackie, A.M. (1979). Cellular recognition of foreign-ness in two insect species, the American cockroach and the desert locust. *Immunology*, **36**, 909–14.

Latgé, J.P., Sampedro, L. & Hall, R (1984). Agressivité de *Conidiobolus obscurus* vis-à-vis du puceron du pois. III Activités enzymatiques exocellulaires. *Entomophaga*, **29**, 185–201.

Lisansky, S.G. & Hall, R.A. (1983). Fungal control of insects. In: *The Filamentous Fungi, Vol. IV, Fungal Technology*, eds J.E. Smith, D.R. Berry & B. Kristiansen, pp. 327–45. Edward Arnold, London.

Locke, M. (1984). The structure of insect cuticle. In: *Infection Processes of Fungi*, a Bellagio Conference, 1983, eds D.W. Roberts & J.R. Aist, pp. 38–53. The Rockefeller Foundation, New York.

Magoon, J. & Al-Aidroos, K.M. (1984). Determination of ploidy of sectors formed by mitotic recombination in *Metarhizium anisopliae*. *Transactions of the British Mycological Society*, **82**, 95–8.

Magoon, J. & Al-Aidroos, K.M. (1986). Epistatic relationships and linkage among colour markers of the imperfect entomopathogenic fungus *Metarhizium anisopliae*. *Canadian Journal of Genetics and Cytology*, **28**, 98–100.

Maiti, I.B. & Kolattukudy,P.E. (1979). Prevention of fungal infection of plants by specific inhibition of cutinase. *Science*, **205**, 507–8.

Messias, C.L. & Azevedo, J.L. (1980). Parasexuality in the Deuteromycte *Metarhizium anisopliae*. *Transactions of the British Mycological Society*, **75**, 473–7.

Messias, C.L., Daoust, R.A. & Roberts, D.W. (1986). Virulence of a natural isolate, auxotrophic mutants, and a diploid of *Metarhizium* var. *anisopliae* to *Rhodnius prolixus*. *Journal of Invertebrate Pathology*, **47**, 231–3.

Messias, C.L., Roberts, D.W. & Grefig, A.T. (1983). Pyrolysis–gas chromatography of the fungus *Metarhizium anisopliae:* an aid to strain identification. *Journal of Invertebrate Pathology*, **42**, 393–6.

Michel, B. (1981). Recherches experimentales sur la pénétration des champignons pathogènes chez les insectes. These 3e cycle, University of Montpellier.

Milner, R.J. & Lutton, G.G. (1986). Dependence of *Verticillium lecanii* (Fungi: Hyphomycetes) on high humidities for infection and sporulation using *Myzus persicae* (Homoptera: Aphididae) as host. *Environmental Entomology*, **15**, 380–2.

Paris, S. & Ferron, P. (1979). Study of the virulence of some mutants of *Beauveria brongniartii* (= *B. tenella*). *Journal of Invertebrate Pathology*, **34**, 71–7.

Paris, S. & Segrétain, G. (1975). Caractères physiologiques de *Beauveria tenella* en rapport avec la virulence des souches de ce champignon pour la larve de Haunneton commun *Melolantha melolantha*. *Entomophaga*, **20**, 135–8.

Paris, S., Ferron, P., Fargues, J. & Robert, P. (1985). Physiological characteristics and virulence of auxotrophic and morphological mutants of *Beauveria brongniartii* (Sacc.) Petch (= *Beauveria tenella*). *Mycopathologia*, **91**, 109–16.

Peberdy, J.F. (1979). Fungal protoplasts: isolation, reversion and fusion. *Annual Review of Microbiology*, **33**, 21–39.

Pekrul, S. & Grula, E.A. (1979). Mode of infection of the corn earworm (*Heliothis zea*) by *Beauveria bassiana* as revealed by scanning electron microscopy. *Journal of Invertebrate Pathology*, **34**, 238–47.

Pontecorvo, G. (1959). *Trends in Genetic Analysis*, Columbia University Press, New York.

Prior, C. & Arura, M. (1985). The infectivity of *Metarhizium anisopliae* to two insect pests of coconuts. *Journal of Invertebrate Pathology*, **45**, 187–94.

Quiot, J.M., Vey, A. & Vago, C. (1985). Effects of mycotoxins on invertebrate cells *in vitro*. In: *Advances in Cell Culture, Vol 4*, ed. K. Maramorosch, pp. 199–212. Academic Press, New York.

Riba, G., Azevedo, J.L., Messias, C., Dias Da Silveira, W. & Tuveson, R. (1985). Studies of the inheritance of virulence in the entomopathogenic fungus *Metarhizium anisopliae*. *Journal of Invertebrate Pathology*, **46**, 20–5.

Riba, G. & Glandard, A. (1980). Mise au point d'un milieu nutritif pour la culture profonde du champignon entomopathogène *Nomuraea rileyi*. *Entomophaga*, **24**, 317–22.

Robert, A. & Al-Aidroos, K. (1985). Acid production by *Metarhizium anisopliae*: effects on virulence against mosquitoes and on detection of *in vitro* amylase, protease and lipase activities. *Journal of Invertebrate Pathology*, **45**, 9–15.

Roberts, D.W. (1981). Toxins. In: *Microbial Control of Pests and Plant Diseases*, ed H.D. Burges, pp. 441–64. Academic Press, New York.

Roberts, D.W. & Humber, R.A. (1981). Entomogenous fungi. In: *Biology of Conidial Fungi, Vol. 2*, eds G.T. Cole & B. Kendrick, pp. 201–36. New York.

Roberts, D.W. & Humber, R.A. (1984). Entomopathogenic fungi. In: *Infection Processes of Fungi*, a Bellagio Conference, 1983, eds D.W. Roberts & J.R. Aist, pp. 1–12. The Rockefeller Foundation, New York.

Silveira, W.D. & Azevedo, J.L. (1984). Isolation of auxotrophic mutants of *Metarhizium anisopliae* by the filtration enrichment technique. *Brazilian Review of Genetics*, **VII** (1), 1–8.

Smith, R.J., Pekrul, S. & Grula, A. (1981). Requirement for sequential enzyme activities for penetration of the integument of the corn earworm *Heliothis zea*. *Journal of Invertebrate Pathology*, **38**, 335–44.

Soderhall, K. & Unestam, T. (1979). Activation of crayfish serum prophenoloxidase in arthropod immunity: the specificity of cell wall glucan activation and activation by purified fungal glycoproteins. *Canadian Journal of Microbiology*, **25**, 404–16.

St. Leger, R.J., Charnley, A.K. & Cooper, R.M. (1986a). Cuticle-degrading enzymes of entomopathogenic fungi: synthesis in culture on cuticle. *Journal of

Invertebrate Pathology, **48**, 85–95.

St. Leger, R.J., Charnley, A.K. & Cooper, R.M. (1987a). Characterization of cuticle-degrading proteases produced by the entomopathogen *Metarhizium anisopliae*. *Archives of Biochemistry and Biophysics*, **253**, 221–32.

St. Leger, R.J., Cooper, R.M. & Charnley, A.K. (1986b). Cuticle-degrading enzymes of entomopathogenic fungi: cuticle degradation *in vitro* by enzymes from entomopathogens. *Journal of Invertebrate Pathology*, **47**, 167–77.

St. Leger, R.J., Cooper, R.M. & Charnley, A.K. (1987b). Production of cuticle degrading enzymes by the entomopathogen *Metarhizium anisopliae* during infection of cuticles from *Calliphora vomitaria* and *Manduca sexta*. *Journal of General Microbiology*, **133**, 1371–82.

St. Leger, R.J., Cooper, R.M. & Charnley, A.K. (1987c). Distribution of chymoelastases and trypsin-like enzymes in five species of entomopathogenic Deuteromycetes. *Archives of Biochemistry and Biophysics*, **258**, 123–31.

St. Leger, R.J., Cooper, R.M. & Charnley, K. (1986c). Cuticle-degrading enzymes of entomopathogenic fungi: regulation of production of chitinolytic enzymes. *Journal of General Microbiology*, **132**, 1509–17.

Suzuki, A., Kawakami, K. & Tamura, S. (1971). Detection of destruxins in silkworm larvae infected with *Metarhizium anisopliae*. *Agricultural and Biological Chemistry*, **35**, 1641–3.

Tinline, R.D. & Macneill, B.H. (1969). Parasexuality in plant pathogenic fungi. *Annual Review of Phytopathology*, **7**, 147–70.

Tinline, R.D. & Noviello, C. (1971). Heterokaryosis in the entomogenous fungus *Metarhizium anisopliae*. *Mycologia*, **63**, 701–12.

Tulloch, M. (1976). The genus *Metarhizium*. *Transactions of the British Mycological Society*, **66**, 407–11.

Typas, M.A. (1983). Heterokaryon compatibility and interspecific hybridisation between *Verticillium albo-atrum* and *V. dahliae* following protoplast fusion and micro-injection. *Journal of General Microbiology*, **129**, 3043–56.

Van Winklehoff, A.J. & McCoy, C.W. (1981). Sporulation of *Hirsutella thompsonii* Fisher in submerged culture. *Society for Invertebrate Pathology XIVth Annual Meeting*, Montana State University, Bozeman, Montana, p. 30.

Vey, A.J. (1984). Cellular antifungal reactions in invertebrates. In: *Infection Processes in Fungi*, a Bellagio Conference, 1983, eds D.W. Roberts & J.R. Aist, pp. 168–74. The Rockefeller Foundation, New York.

Woods, S.P. & Grula, E.A. (1984). Utilisable surface nutrients on *Heliothis zea* available for growth of *Beauveria bassiana*. *Journal of Invertebrate Pathology*, **43**, 259–69.

Yoder, O.C., Weltring, K., Turgeon, B.G., Garber, R.C. & Van Etten, H.D. (1986). Technology for molecular cloning of fungal virulence genes. In: *Biology and Molecular Biology of Plant–Pathogen Interactions*. ed. J. Bailey, pp. 371–84. Plenum Publishing Corporation, New York.

The use of fungi in integrated control of plant diseases

12.1 Introduction

To begin this discussion, it is necessary to define the major terms inherent in the concept of integrated control. The term 'integrated control' had its origins in entomology. Stern *et al.* (1959) defined IC as 'applied pest control which combines and integrates chemical and biological methods'. With regard to plant pathology, Andrews (1983) suggested that the scope of integrated control is to integrate cultural, biological and chemical strategies for control of plant diseases. Integrated control is, therefore, a flexible, multidimensional approach utilising a range of control components as required to hold diseases below damaging economic levels without disrupting the agroecosystem. Integrated control should not be confused with the broader concept of integrated pest management (IPM), which includes integrated control programmes for various kinds of pests and additional elements including societal values, impact on other resources, and disciplinary and methodological integration.

Since the emphasis in this book is on fungi in biocontrol systems, integrated control in this chapter is no more than a blending of the use of such fungi with other control strategies such as resistant cultivars, pesticides, and cultural measures. There are several narrow and broad definitions for the term 'biocontrol'. To avoid any strenuous and futile exercises in semantics in the present review, we propose to accept the definition of biocontrol suggested by an *ad hoc* group of scientists at the 1983 National Interdisciplinary Biological Control Conference in Las Vegas, Nevada (Baker, 1983). This definition is based on 'pest suppression with biotic agents, excluding the process of breeding for resistance to pests, sterility techniques, and chemicals modifying pest behavior'. The definition is closer to that suggested by Garrett (1965, 1970) than to the broad definition of biocontrol suggested elsewhere (Baker & Cook, 1974; Cook & Baker, 1983).

This review is primarily concerned with integrated control procedures which combine cultural and chemical control approaches with a stimulation of indigenous potential biocontrol fungi or of introduced biocontrol fungi. This paper also discusses the potential for integrated control through development of new pesticide-resistant strains of biocontrol fungi by genetic manipulation. Although every attempt has been made to include field applications of integrated control, information from glasshouse studies is also presented.

12.2 Integrated control with cultural practices, pesticides and indigenous fungi

Development of integrated control strategies depends on identification and integration of appropriate control objectives. The major objectives are: (1) reduction in the populations of disease-causing microorganisms to levels that do not limit crop production, and (2) utilisation of more than one disease-control component in reducing a disease below the economic-injury threshold. In this review, by definition, biocontrol fungi (indigenous or introduced) must be one of the integrated control components.

The use of indigenous fungi in integrated control for disease reduction offers unusual opportunities for integration with other disease management systems because, simply, the fungi are already there. Success in such integrated control systems, however, will depend on specific ecological conditions and on finding the proper selective treatment to enhance the activity and multiplication of antagonistic fungi in the environment (Backman & Rodriguez-Kabana, 1977). The selective treatment may be a cultural practice or a chemical application. The microbiological balance of the soil, often referred to as a 'dynamic equilibrium', is subject to at least temporary alterations, imposed by the selective treatment. Partial or complete soil disinfestation to reduce various pests or weeds — practices that may result in a differential removal of various components of the soil microbiota — will undoubtedly create a soil 'biological vacuum' (Baker, 1981) the magnitude of which would depend on the degree of disinfestation and on the kinds of materials used in the process.

The degree and kind of soil reinfestation by surviving indigenous or introduced microbes appears to be proportional to the degree of reduction of the native microbiota and to certain nutritional changes that occur. As a result of soil treatment, certain soil-borne plant diseases may either increase or decrease (Elad *et al.*, 1982). If the alteration of the microbiological balance, temporary as it may seem to be, is in favour of certain indigenous fungi antagonistic to plant pathogens by virtue of their ability to colonise the soil and fill the vacuum, diseases will decrease. In contrast, if a plant pathogen is the component of the soil mcirobiota that will occupy the soil or fill the vacuum, disease will increase.

12.2.1 *Cultural practices and indigenous fungi*
Cropping systems and other cultural practices are the most important factors in the agroecosystem which determine the activities of the indigenous soil fungi (Baker & Cook, 1974). Crop rotation and tillage practices can reduce or intensify some diseases by altering the ecosystem to stimulate or suppress a pathogen. However, it has generally not been possible to establish whether the effect is direct or indirect. Soil heating is perhaps the only cultural or physical treatment that has been intentionally incorporated into integrated control systems (Sewell, 1965).

Sub-lethal heating of soil infested with *Armillaria mellea*, a fungus that attacks a variety of tree roots, weakened the pathogen and stimulated the more heat-resistant indigenous *Trichoderma*, an important biocontrol fungus (Papavizas, 1985), which then proliferated and parasitised the weakened *A. mellea* (Munnecke *et al.*, 1981). The intensity and duration of the heat applied did not kill the citrus plants or the pathogen but reduced survival of *A. mellea*. Treatment of soil with aerated steam has been used since 1888 to kill or weaken soil-borne plant pathogens. However, in this process, a considerable portion of the indigenous microflora is also destroyed, creating a situation which allows surviving microorganisms to proliferate. Many of these survivors include fungi (especially of the genera

Aspergillus, Gliocladium, Penicillium and *Trichoderma*) which can reduce disease potential further by killing weakened pathogens and preventing reinfestation (Bollen, 1969; Broadbent *et al.*, 1971; Baker & Cook, 1974).

Heating soil on a large field scale can also be accomplished by a process called solarisation in which heat from the sun penetrates a plastic mulch placed on moist soil and essentially 'disinfects' the soil (Katan, 1981). Solarisation can control disease by direct destruction of the pathogen or by weakening the pathogen to such an extent that it is attacked by the micro-flora which survives the solarisation. Heating by solarisation has been accountable for the destruction of *Sclerotium rolfsii* and *Fusarium* in soil in which antagonistic indigenous *Trichoderma* populations increased (Elad *et al.*, 1980).

12.2.2 *Pesticides, chemicals and indigenous fungi*
12.2.2.1 *Soil-borne diseases.* Before the early 1970s, much of the research on biocontrol focused on the indirect enhancement of indigenous microorganisms, especially *Trichoderma* (Papavizas, 1985), by manipulating the soil environment with pesticides. Yet the precise role of these indigenous populations in the biocontrol of diseases is generally only a matter of speculation. Few scientists have discussed the significance of combining the two control components, pesticides and biocontrol, at least not in the context of integrated control.

Over the years, considerable attention has focused on the interaction between various soil fumigants and indigenous potential biocontrol fungi. The classical example of this type of interaction, demonstrated 35 years ago by Bliss (1951), is the control of *Armillaria mellea* by carbon disulphide treatment of citrus-orchard soil. Bliss attributed the control following the fumigation to the action of the indigenous *Trichoderma* that rapidly colonised and reproduced in the fumigated soil, but he did not provide direct evidence that *Trichoderma* controlled the pathogen. More recently, sub-lethal concentrations of methyl bromide reduced *A. mellea* in soil and the reduction was attributed to the fact that *Trichoderma* was more resistant to methyl bromide than *A. mellea* (Munnecke *et al.*, 1981). *A. mellea*, weakened by the fumigant, produced less of the antibiotic that otherwise protected it against *Trichoderma* in non-fumigated soil. A complete discussion of the *A. mellea*–fumigant interaction can be found elsewhere (Munnecke *et al.*, 1981).

In several other studies, fumigants have been implicated in causing an increase in population size or activity of indigenous soil fungi. For example, the addition of sodium azide to soil increased the populations of *Trichoderma* (Kelley & Rodriguez-Kabana, 1978). Similar increases were demonstrated with the use of carbon disulphide (Moubasher, 1963) and field application of methyl bromide (Strashnow *et al.*, 1985). Strashnow *et al.* (1985) postulated that the long-range beneficial effect of the chemical treatment in disease reduction was due not to the direct toxicity of the pesticide to any particular pathogen, but to a stimulation of antagonists that eliminated the pathogen. Treatment of soils with the fumigants allyl alcohol, chloropicrin, dichloropropenes and methyl bromide in the laboratory promoted the development of dominant populations of *Trichoderma* which comprised almost 100% of the recolonising fungus flora (Mughogho, 1968). Mende (cited by Bochow, 1975) tested the effects of limited amounts of metham and dazomet for soil fumigation of light soils in vegetable production under glass and plastic cover. The pesticide applications improved yield and performance of cucumber, lettuce and tomato. He considered these results as a proof of synergism between the pesticide and indigenous beneficial soil fungi under commercial conditions and as an example of integrated control.

Fungicides other than fumigants have been used as components in the integrated control of plant diseases, although the scientists that used them generally did not have integrated control in mind at the time nor did they intend to increase indigenous soil fungi. The addition of sub-lethal amounts of methyl mercury dicyandiamide to *Helminthosporium sativum*-infested soil reduced common root rot of wheat, but did not affect the pathogen (Chinn, 1971). The treatment increased numbers of soil fungi, especially potentially antagonistic species of *Penicillium*. In studies on narcissus bulb rot caused by *Fusarium oxysporum* f. sp. *narcissi*, the fungicides thiram, organic mercury and formalin and the antibiotic pimaricin were used as bulb treatments. In addition to disinfecting the bulbs, the chemicals protected the roots even though the residue on the bulbs decreased below a level toxic to the pathogen (Langerak, 1977). The narcissus roots were colonised by species of *Penicillium* and *Trichoderma*, which were stimulated by the chemicals and were antagonistic to the pathogen. Thiram effectiveness against *Pythium* damping-off of peas was evident long after the chemical had decomposed in soil because it had stimulated populations of saprophytic fungi, particularly *Trichoderma* and *Penicillium*, which were antagonistic to the pathogen (Richardson, 1973). Field application of captan, dicloran, milcol, and triarimol reduced the number of fungi but after several weeks *Trichoderma* and *Penicillium* became dominant in the treated soils (Wainwright & Pugh, 1975). Additional aspects of pesticide-induced biocontrol were discussed by McKeen (1975).

Fungicides and fumigants are not the only chemicals that have been used to increase activities and induce proliferation of soil fungi. Although short- and long-term effects of insecticides, nematicides and herbicides on non-target fungi have been reviewed extensively (Katan & Eshel, 1973; Altman & Campbell, 1977; Papavizas & Lewis, 1979), very little information is available on the integrated control of plant diseases with combinations of native fungi and insecticides or herbicides; and the small amount of information found is conjectural at best. The nematicides etho-prop and fensulfothion, for instance, inhibited growth and production of sclerotia of *Sclerotium rolfsii in vitro* without adversely affecting some of the dominant fungal genera (Rodriguez-Kabana *et al.*, 1976a, b). In fact, the two nematicides stimulated growth and proliferation of *Trichoderma* and *Aspergillus* in soil and indirectly enhanced invasion of *S. rolfsii* colonies by *Trichoderma*. Rodriguez-Kabana *et al.* (1976a, b) also obtained limited field evidence that the mycoparasitic action of *Trichoderma*, coupled with the fungistatic ability of the two nematicides on *S. rolfsii*, reduced peanut stem blight.

Fertilisers have also been implicated in stimulating indigenous soil fungi responsible for biocontrol. For example, sulphur added to soil to maintain the pH below 3.9 reduced root rot and heart rot of pineapple in Australia (Cook & Baker, 1983). The control was attributed to a decrease in zoo-sporangium formation of the causal agent (*Phytophthora cinnamomi*) and to an increase of the acidophilic native *T. viride*. Also, the mechanism of pine stump protection against *Heterobasidion annosum* provided by urea or disodium octaborate used in England (Garrett, 1965) or by sodium nitrite or borax used in the United States (Driver, 1963) seems to operate chiefly through phytocidal effects of such chemicals, which open the way for biocontrol of *H. annosum* by various competing fungal saprophytes.

12.2.2.2 *Foliar diseases.* Reports on the integrated control of foliar and other above-ground diseases involving indigenous biocontrol fungi and pesticides or miscellaneous chemicals are rarely found in the literature. However, one such case concerns the prevention of a severe above-ground disease of cyclamen caused by benomyl-resistant strains of *Botrytis cinerea* by the combined effects of benomyl and the epiphytic microflora, especially species of *Penicillium* that became resistant to the fungicide (van Dommelen & Bollen, 1973). Benomyl induced resistance in both *B. cinerea* and *Penicillium*. For other foliar diseases, the effects of more subtle interactions, such as those involving epiphytic microflora and fungicides, are just beginning to be identified.

Integration of indigenous biocontrol fungi with other control components would in all probability enhance the prospects for single or multiple plant disease control over the use of individual control components. Such a dual system, taking advantage of the existing beneficial fungi in the ecosystem, would undoubtedly be less costly than a system involving introduced biocontrol fungi. This system assumes that stable populations of antagonistic fungi would be established in pesticide-treated soils or soils exposed to various cropping systems. Integrated control based on the indigenous fungal flora would only be feasible if the antagonistic fungi were compatible with other control practices.

12.3 Interactions of pesticides, fertilisers and cultural practices with introduced biocontrol fungi

Several important economic diseases occur on our crop plants that provide food and fibre. Applications of fungicides and fumigants to soil, seed or foliage are approaches used for disease management. Several political, environmental and economic deterrents are encountered in the use of pesticides in the United States and in many other countries. These deterrents may be tempered somewhat by finding integrated uses of pesticides at reduced rates together with biocontrol fungi and other agents to suppress single diseases or combinations of diseases and other pests. In many instances, indigenous fungal microflora including antagonists may be eliminated or unable to proliferate following an alteration of the ecosystem with selective pesticide treatments. In such instances, the biological vacuum may be filled by introduced pesticide-resistant antagonistic fungi. Integration of pesticide-tolerant antagonistic fungi with chemical controls of plant diseases would render the practice more effective for multiple disease control over the use of only one control practice. In such cases, it is preferable not to introduce antagonistic fungi that possess only a high survival ability of a passive nature; it is desirable to introduce fungi with a high inherent ability to utilise liberated nutrients in soil and to recolonise the soil. It is also desirable to select a biocontrol agent of multiple disease control potential that may be integrated with other biological agents, pesticides or other agronomic practices (Papavizas, 1973; Klassen, 1981; Smith, 1982).

In the early research on soil disinfestation and sterilisation, it was observed that the early, aggressive, visible colonisers of the treated soils were fungi of the genera *Trichoderma* and *Gliocladium* (Papavizas, 1985). The dominance of these two genera in subsequent studies with introduced biocontrol fungi was based on their ability to proliferate in the treated environment as well as their demonstrated ability to produce antibiotics, to act as mycoparasites, and to compete for nutrients. Consequently, a majority of the reported research activities with introduced biocontrol fungi are with isolates of *Trichoderma* and *Gliocladium*, fast-growing organisms which develop on a variety of substrates and are easily handled.

12.3.1 *Pesticides and introduced fungi*
12.3.1.1 *Soil-borne diseases.* Research and technology on integrated control systems involving pesticide application and introduction of natural fungi or new fungal biotypes resistant to pesticides has been limited. However, during the past 10 to 15 years this concept has been advanced in attempts to reduce various diseases. Some positive results obtained in the glasshouse provided the impetus for additional field testing. Judging from the literature of the last few years, there has been considerable interest in

combining biocontrol fungi with fumigants and other fungicides. This interest is reflected in the increasing numbers of publications and reviews, especially on the integrated control of weeds (Smith, 1982). Even though information on this subject is accumulating rapidly, very few definite statements can be made about the commercial applications in the field. The main deterrent to the integration of commercially used pesticides with biocontrol fungi is the lack of commercial formulations of registered biocontrol fungi. Some of the reasons for this deterrent were discussed elsewhere (Papavizas & Lewis, 1981; Papavizas, 1985).

As indicated earlier, biocides can stress and weaken pathogen & propagules and render them more susceptible to attack by antagonistic fungi. Sclerotia of *Sclerotium rolfsii* 'weakened' by sub-lethal concentrations of metham became susceptible to invasion and degradation by introduced *Trichoderma harzianum* (Henis & Papavizas, 1983); *T. harzianum* alone did not degrade fresh viable sclerotia. In field studies with fumigants, methyl bromide is the chief material used in association with biocontrol fungi. A combination of methyl bromide and *T. harzianum* controlled *Rhizoctonia solani* and *S. rolfsii* on tomatoes, peanuts and strawberries (Elad *et al.*, 1981, 1982). Although a reduced rate of application of the fumigant did not control *R. solani* on carrots, application of the sub-lethal rate to soil with a bran–peat preparation of *T. harzianum* was as effective as the recommended rate of the fumigant for disease reduction (Strashnow *et al.*, 1985). The antagonist *Myrothecium verrucaria*, effective against *Phytophthora cinnamomi* on avocado, was grown on grain and applied to infested soil after methyl bromide fumigation (Munnecke, 1984). The biocontrol fungus in combination with the fumigant delayed disease development and reduced severity in the field but did not prevent it. In addition, *M. verrucaria* populations were established in the field for 5–7 months.

Other major field studies with fumigants and introduced biocontrol fungi have been reported. Crown rot of tomatoes caused by *Fusarium oxysporum* f. sp. *radicis-lycopersici* was reduced in the field after introduction of conidia of *Aspergillus*, *Penicillium* or *Trichoderma* into methyl bromide and chloropicrin fumigated soils (Marois *et al.*, 1981). Recent research at Beltsville has suggested the potential of combining pesticides with introduced biocontrol fungi for control of diseases in the field caused by soilborne plant pathogens. A good example of this is the successful biocontrol of eggplant wilt caused by *Verticillium dahliae* in production systems by the introduction of *Talaromyces flavus* (anamorph: *Penicillium dangeardii*), an Ascomycete isolated from decaying sclerotia of *Sclerotinia minor* found in natural soil, to fields treated with the pesticide dichloropropene + methylisothiocyanate (one-quarter of the normal rate) (Marois *et al.*, 1982). Ascospores or conidia of the biocontrol fungus were applied to the transplanting mix in the glasshouse before moving the seedlings to the fumigated field. The biocontrol fungus reduced *Verticillium* wilt by about 70% and increased yield by 54 and 71% in two fields. When applied together with the reduced amount of the fumigant, *T. flavus* gave control equivalent to that of the full rate of the fumigant. *Talaromyces flavus* also suppressed *Verticillium* wilt of potato in 2 years of field tests in Idaho in metham-fumigated soil (Davis *et al.*, 1986; Fravel *et al.*, 1986). The best reduction of wilt was obtained with a triple treatment involving *T. flavus* ascospores that were applied to thiabendazole-treated potato seedpieces which were then planted in metham-fumigated soils.

With regard to other fungicides, and introduced biocontrol fungi, Davet & Martin (1985) found that colonisation and destruction of sclerotia of *S. minor* by a *Trichoderma* sp. was enhanced when cyclic imides were also added to the *Trichoderma*-enriched soils. *Trichoderma* tolerates the imides. Chet *et al.* (1979) integrated the use of *T. harzianum* and low dosages of pentachloronitrobenzene (PCNB) to manage damping-off disease of bean seedling caused by *S. rolfsii*. Others (Henis *et al.*, 1978; Hadar *et al.*, 1979) integrated the same two components to manage damping-off of eggplant and radish seedlings caused by *R. solani*. When Henis *et al.* (1978) applied wheat-bran cultures of *T. harzianum* to soil infested with the pathogen, radish seedlings were protected from the disease. The fungicide PCNB, added at 4 µg active ingredient per gram of soil with *T. harzianum*, had an additive effect on disease control and a synergistic effect on the decrease of inoculum density of *R. solani* propagules. Similarly, application of a *T. harzianum* bran culture in association with prothiocarb to a commercial nursery rooting mixture infested with *R. solani* and *Pythium aphanider-matum* prevented damping-off of gypsophila cuttings better than either of the single components (Sivan *et al.*, 1984). Other examples of a pesticide and an introduced biocontrol agent include combining *T. harzianum* and PCNB against *Rhizoctonia* damping-off of vegetables (Elad *et al.*, 1980) and bulb-borne pathogens of iris (Chet *et al.*, 1982), and the combination of *T. hamatum* and the fungicide metalaxyl against *Pythium* blight on turf (Rasmussen-Dykes & Brown, 1982).

12.3.1.2 *Foliar diseases.* Attempts have also been made to control foliar diseases using sprays with fungicides in combination with compatible bio-control fungi. Since the fungi are sprayed on crops along with or sub-sequent to the pesticide, the fungal preparations must be suitable for spray equipment. Consequently, most research is currently performed with spores prepared from solid media or homogenised cultures. Commercial applica-tion based on preparations containing spores may not always be feasible. However, spray preparations of finely powdered fermenter biomass con-taining the effective propagules may be viable alternatives (Papavizas *et al.*, 1984). In Italy (Lantero *et al.*, 1982) and France (Dubos *et al.*, 1982),

Botrytis cinerea on grapes was controlled with sprays done at flowering with homogenised cultures of *T. viride* in association with the fungicides vinclozolin or dichlofluanid. With this approach, less chemical and fewer sprays were required for adequate disease control.

The feasibility of integrated control with a biocontrol fungus and a pesticide was shown in studies designed to find ways to reduce damage from the dieback disease of apricot caused by *Eutypa armeniacae*. Carter (1971) found that a suspension of spores of *Fusarium lateritium* sprayed on apricot trees within 24 h after pruning provided some protection from infection by *E. armeniacae*. The fungicide benomyl was effective against the pathogen, but not against *F. lateritium*. The combined use of benomyl and *F. lateritium* gave better protection against *E. armeniacae* than when either component was used alone. The integrated control was possible because, fortuitously, *F. lateritium* was ten times less sensitive to benomyl than the pathogen (Carter & Price, 1974, 1975). In a later study, good disease protection of the pruned apricot sapwood was obtained by combining the fungicide and the antagonist in the field.

Success in integrated control involving the combined use of pesticides and biocontrol fungi would depend not only on disease reduction, but also on yield increases. *Sphaerotheca fuligena*, foliar pathogen that causes a serious mildew disease on cucumber, is controlled with the fungicide triforine and is also attacked by the hyperarasitic fungus *Ampelomyces quisqualis* (Sundheim & Amundsen, 1982). The hyperparasite alone does not give full protection against the pathogen. When plants were sprayed with reduced rates of triforine (one-third of the recommended rate) alternating with sprays of spore suspensions of the hyperparasite, which is naturally tolerant to the fungicide, the yield increased 50% over that of the controls. The integrated control resulted in even higher yields than those produced by the normal rate of triforine.

Attempts to suppress diseases combining fungicides and biocontrol fungi have not always been successful and negative results are usually not reported. For example, methyl bromide and *T. harzianum* failed to reduce potato diseases or increase yields over the fumigant used alone (Elad *et al.*, 1980). The control of stem rot of peanuts caused by *S. rolfsii* with the fungicides PCNB and carboxin was not improved by application of preparations of *T. harzianum* with the fungicides (Csinos *et al.*, 1983). Wheat bran preparations of several potential biocontrol fungi did not significantly improve the efficacy of metalaxyl seed treatment for the control of *Rhizoctonia* and *Pythium* diseases of snap beans in the field (Lewis *et al.*, 1983).

The use of tolerant strains of biocontrol fungi together with selective pesticides offers distinct possibilities of plant disease control in cases where no single component is effective.

12.3.2 *Cultural practices and introduced fungi*

The development of field technology for efficient and reliable integrated management of plant diseases with cultural practices and antagonistic fungi, or other microbes, is lagging behind that for management of insects (Papavizas, 1973). Crop rotation, for instance, is a valuable control component in pest management, as shown by the impressive list of examples where the sequence of crops in rotation has reduced crop damage from certain soil-borne plant pathogens (Lumsden *et al.*, 1983). Although crop rotation has been directly associated with the build-up in soil of indigenous fungi, no published reports are available on integration of introduced fungi and crop rotation or tillage systems. Very few reports can be found on integrated control involving introduced fungi and other cultural practices such as soil cultivation, solarisation, or fertiliser application.

In warm humid areas, fruit rot (soil rot), caused by *Rhizoctonia solani*, is a serious disease causing heavy losses on cucumber and tomato. The disease is increased by monoculture and high-density planting needed for mechanical harvesting. Lewis & Papavizas (1980) showed that the fruit rot of cucumber could be reduced by one-half by ploughing infested fields 20–25 cm deep with a mouldbroad plough rather than discing the fields 4–6 cm deep before planting. They also showed that the use of the fungi *Laetisaria arvalis* (*Corticium* sp.) and *Trichoderma harzianum* could be effectively integrated with mechanical ploughing to suppress the disease. The use of these fungi in association with ploughing resulted in less disease than when either component was used individually. The control obtained was as good as that obtained with fungicides registered for control of fruit rot on cucumber. This study is one of very few documented examples integrating antagonistic fungi with cultural methods against plant pathogens.

Soil solarisation is another interesting approach for physical manipulation of the soil to be used in association with introduced biocontrol fungi. Bran–sawdust preparations of *T. harzianum* applied to soil after solarisation did not significantly increase effectiveness of solarisation in reducing potato diseases caused by *R. solani*, *S. rolfsii* and *V. dahliae* (Elad *et al.*, 1980). However, the combination reduced the inoculum potential of *R. solani* and its proliferation in the field.

Steam sterilisation or pasteurisation can also be used as a selective physical treatment in an integrated system with introduced fungi. Excellent control of *Fusarium* wilt of chrysanthemum was obtained by the addition to soil mix of aqueous conidial suspensions of a benomyl-resistant biotype of *Trichoderma viride* (Locke *et al.*, 1985). The biotype, used at 10^4 conidia/cm^2 after the soil mix was pasteurised with steam (at 82°C for 2 h), rapidly colonised the soil mix (Marois & Locke, 1985), prevented reinvasion by the pathogen, and provided control equal to a commercial integrated procedure that involves benomyl and fertilisers, but not biocontrol agents.

The benomyl-resistant biotype was not inhibitory to *Fusarium oxysporum* f. sp. *chrysanthemi* in cultures (Papavizas & Lewis, 1983).

One or more biocontrol fungi may also be combined with fertilisers to control a disease or used sequentially to control several diseases. Unfortunately, no research has been done on such combinations in the field. Glasshouse studies showed that urea fertiliser combined with *T. harzianum* grown on a wheat-bran formulation reduced the viability of sclerotia of *Sclerotium rolfsii* in soil more effectively than when either component was used alone (Maiti & Sen, 1985). Calcium ammonium nitrate and *T. harzianum* were most effective in reducing stem blight of groundnuts caused by *S. rolfsii*. The results of this kind of integrated control, however, were not verified in production systems. Gypsum, applied to the field, reduced aflatoxin production in peanuts caused by *Aspergillus flavus* whereas inoculum of *T. harzianum* grown on wheat grain did not (Mixon *et al.*, 1984). When the two components were combined there was generally less pod colonisation by *A. flavus* than when gypsum was used alone.

12.3.3 *Resistant cultivars and introduced fungi*

The single published example available in the literature on the combined use of introduced biocontrol fungi and cultivar resistance against a plant pathogen is that published by Kraft & Papavizas (1983). They described the results of an integrated program used in the pea area of Washington State near Prosser to reduce losses from damping-off of peas caused by *Pythium ultimum* and *Fusarium solani* f. sp. *pisi* and to increase yields. In this system, the highest seed yields with the susceptible cultivar, Dark Skin Perfection, were obtained with a seed treatment combining metalaxyl and *T. harzianum* conidia. No benefits were apparent from treating seed of three resistant breeding lines. This was in contrast to the response of Dark Skin Perfection, where treating seed with *T. harzianum* resulted in significant increases in plant stand and yields compared with the untreated control.

12.4 **Selection and production of new strains compatible with pesticides**

One of the main deterrents to the use of biocontrol fungi in integrated control systems together with pesticides is their sensitivity to such chemicals. For instance, the mycoherbicide *Colletotrichum gloeosporioides* f. sp. *aeschynomene* failed to control the weed northern jointvetch when spore suspensions of the fungus were unintentionally combined with benomyl before aerial application (Smith, 1982). If fungi are to be integrated effectively with pesticides to control single or multiple diseases, they must have natural resistance to pesticides, or be manipulated genetically to develop such resistance.

Attempts at realisation of practical integrated approaches cannot and should not always be passive from the standpoint of selecting biocontrol fungi tolerant to fungicides. The few advances in integrated control described so far have been accomplished with biocontrol fungi which, fortuitously, were tolerant to certain pesticides. From a dynamic stand-point, there are various options available to establish an effective scheme that will encompass biological and chemical control approaches for single of multiple pest control. The first option involves active selection of antagonistic fungi that may tolerate pesticides. Uyttebroeck *et al.* (1979) tested several isolates of *Tichoderma* spp. in the laboratory for their ability to tolerate methyl bromide. The isolates differed in their sensitivity to the fumigant. A few selected isolates obtained from soil treated with the fumigant exhibited strong antagonistic activity *in vitro* against *Rhizoctonia solani*. In a similar effort but with a broader approach, Baicu (1982) studied toxicity of several fungicides, acaricides and insecticides to the growth and spore germination of *T. viride* and the effects of such pesticides on the antagonistic activity of *T. viride* to *Fusarium graminearum*. Benomyl and thiophanate methyl were the compounds most toxic to the antagonist. A few insecticides and the fungicides folpet, wettable sulphur, quinomethioate, dinocap and copper oxychloride were non-toxic to mycelial growth and germination of conidia.

Lewis & Papavizas (1984) studied the effect of the fumigant metham on 34 isolates of *T. hamatum*, *T. harzianum* and *T. viride* in order to determine the fate of potential biocontrol isolates when introduced into soils before, during or after treatment with sublethal rates of the fumigant. Although small amounts of the isothiocyanate released from the metham in soil inhibited fungal growth and germination of conidia, the most encouraging finding from the standpoint of integrated control was the observation that exposure of spores to metham in solution for 1 h allowed considerable germination of conidia of the three fungi. Obviously, then, conidia of *Trichoderma* spp. may be applied in tank mixtures or 'chemigation' systems together with metham or perhaps other pesticides, or coated on to seeds together with infused seed treatment fungicides (Papavizas, 1981).

Some of the fungi commonly used for biocontrol possess natural resistance to several commercially used fungicides and fumigants. *Trichoderma* spp., for instance, tolerated several fungicides including chloroneb, captan, PCNB and metalaxyl (Abd-El Moity *et al.*, 1982). Some of these fungicides are used in selective media to isolate *Trichoderma* and *Gliocladium* from soil (Papavizas & Lumsden, 1982). Infusion of pea seed with the fungicide metalaxyl before coating it with conidia of *T. harzianum* improved survival of conidia in the rhizosphere compared with the survival in the rhizosphere from seed that received conidia only (Papavizas, 1981). Metalaxyl is even used as a carbon source by *Trichoderma* (G.C. Papavizas, unpublished).

Wild-type isolates of *Talaromyces flavus* were shown to have a natural resistance to the fungicides thiabendazole and metiram recommended for use on potato seedpieces (Fravel *et al.*, 1985). However, not all biocontrol fungi are resistant to commonly used fungicides, and almost all of them are very sensitive to the methyl benzimidazole carbamate (MBC) group of fungicides.

There is now considerable interest in applying genome modifications to improve the efficacy of biocontrol fungi and their resistance to pesticides. The exciting idea of tailoring fungal antagonists that will combine the traits of effective biocontrol capabilities and decreasing sensitivity to pesticides, especially fungicides, is no longer far-fetched. The trait of biocontrol capability, probably determined polygenically in fungi, will be more difficult to improve than the trait of pesticide resistance usually governed by a single gene. Resistant strains of biocontrol fungi can be obtained by mutagenesis, conventional genetic approaches, or genetic engineering. Considerable progress has already been made in developing biocontrol agents (*Trichoderma, Talaromyces*) resistant to MBC fungicides by physical and chemical mutagenesis (Papavizas *et al.*, 1982; Papavizas & Lewis, 1983; Papavizas, 1986) and by sexual hybridisation (Katan *et al.*, 1984). Benomyl resistance was also chemically induced in *Chaetomium globosum*, a fungus which attacks the apple scab pathogen *Venturia inaequalis* (Cullen & Andrews, 1985). The wild strain reduced disease afer foliar application of ascospores but could not tolerate the benomyl used in routine spraying. There was significantly less disease in the field when benomyl was used together with the resistant biotype than with either individual component.

No progress has been reported yet in developing biocontrol fungi resistant to pesticides by molecular genetics and genetic engineering, although strains of *Aspergillus nidulans*, not known as a biocontrol agent, have been constructed that are resistant to the antibiotics ampicillin and chloramphenicol (Yelton *et al.*, 1984).

Since genetic manipulations and molecular genetics to develop resistance to pesticides and to improve biocontrol capabilities have been reviewed recently (Papavizas, 1985, 1986), there is no need to discuss this subject further. It will suffice to say only that purposeful induction of tolerance or resistance to pesticides in biocontrol fungi and the selection of stable biotypes for use in combination with pesticides are one of several options available for establishing an effective control scheme that will encompass biological and chemical approaches to single or multiple pest control.

12.5 Conclusions

Plant pathologists and soil microbiologists, in immersing themselves in the intriguing prospects of biocontrol *per se* and recently in the challenging and exciting molecular and biotechnological approaches of pest control, have practically abandoned the integrated control concept as rapidly as they adopted it a few years ago, thus neglecting the possibilities of an integrated control of plant diseases with biocontrol as one of its components. However, research results, some of which we have mentioned, clearly indicate that integrated control utilising biocontrol fungi as a component is a viable as well as an intriguing prospect. It appears that the number of field tests based on glasshouse observations has increased. Although a considerable number of these experiments may not have been entirely successful, information has accumulated to enable the design and perfection of future tests. The concept of integrated control is feasible and agriculturally acceptable. Rather than neglect integrated control, it is necessary to continue research in this area utilising the new biotechnological advances (e.g. genome modification) which are becoming increasing available.

The future for integrated control in plant pathology was discussed recently by Andrews (1983) who stated that

> ...integrated control needs to be a major goal for pathologists in the coming decades. An outstanding base of fundamental and applied information exists for this endeavor. The future strategies that are developed will have major implications not only for disease control, but for the composition of IPM programs and the management of agroecosystems.

References

Abd-El Moity, T.H., Papavizas, G.C. & Shatla, M.N. (1982). Induction of new isolates of *Trichoderma harzianum* tolerant to fungicides and their experimental use for control of white rot of onion. *Phytopathology*, **72**, 396–400.

Altman, J. & Campbell, C.L. (1977). Effect of herbicides on plant diseases. *Annual Review of Phytopathology*, **15**, 361–85.

Andrews, J.H. (1983). Future strategies for integrated control. In: *Challenging Problems in Plant Health*, eds T. Kommedahl & P.H. Williams, pp. 431–40. American Phytopathological Society, St. Paul, Minn.

Backman, P.A. & Rodriguez-Kabana, R. (1977). Predisposition of peanuts to disease and suppression of *Sclerotium rolfsii* by pesticides: the role of antagonists. In: *Current Topics in Plant Pathology*, ed. Z. Kiraly, pp. 209–14. Z. Akad. Kaido, Budapest.

Baicu, T. (1982). Toxicity of some pesticides to *Trichoderma viride* Pers. *Crop Protection*, **1**, 349–58.

Baker, K.F. (1981). Biological control. In: *Fungal Wilt Diseases of Plants*, eds M.E. Mace, A.A. Bell & C.H. Beckman, pp. 523–61. Academic Press, New York.

Baker, K.F. & Cook, R.J. (1974). *Biological Control of Plant Pathogens*. W.H. Freeman & Company, San Francisco, CA.

Baker, R. (1983). State of the art: plant diseases. In: *Proceedings of the National Interdisciplinary Biological Control Conference*, ed. S.L. Battenfield, pp. 14–22. CSRS/USDA, Washington, DC.

Bliss, D.E. (1951). The destruction of *Armillaria mellea* in citrus soils. *Phytopathology*, **41**, 665–83.

Bochow, H. (1975). Possibilities and problems of an integrated plant protection programme in vegetable production. *Proceedings of the International Horticultural Congress*, **19**, 395–408.

Bollen, G.J. (1969). The selective effect of heat treatment on the microflora of a greenhouse soil. *Netherlands Journal of Plant Pathology*, **75**, 157–63.

Broadbent, P., Baker, K.F. & Waterworth, Y. (1971). Bacteria and actinomycetes antagonistic to fungal root pathogens in Australian soils. *Australian Journal of Biological Sciences*, **24**, 925–44.

Carter, M.V. (1971). Biological control of *Eutypa armeniacae*. *Australian Journal of Experimental Agriculture and Animal Husbandry*, **11**, 687–92.

Carter, M.V. & Price, T.V. (1974). Biological control of *Eutypa armeniacae*. II. Studies of the interaction between *E. armeniacae* and *Fusarium lateritium*, and their relative sensitivities to benzimidazole chemicals. *Australian Journal of Agricultural Research*, **25**, 105–19.

Carter, M.V. & Price, T.V. (1975). Biological control of *Eutypa armeniacae*. III A comparison of chemical, biological and integrated control. *Australian Journal of Agricultural Research*, **26**, 537–43.

Chet, I., Elad, Y., Kalfon, A., Hadar, Y. & Katan, J. (1982). Integrated control of soilborne and bulbborne pathogens in iris. *Phytoparasitica*, **10**, 229–36.

Chet, I., Hadar, Y., Elad, Y., Katan, J. & Henis, Y. (1979). Biological control of soil-borne plant pathogens by *Trichoderma harzianum*. In: *Soil-borne Plant Pathogens*, eds B. Schippers & W. Gams, pp. 585–91. Academic Press, London.

Chinn, S.H.F. (1971). Biological effect of panogen PX in soil on common root rot and growth response of wheat seedlings. *Phytopathology*, **61**, 98–101.

Cook, R.J. & Baker, K.F. (1983). *The Nature and Practice of Biological Control of Plant Pathogens*. The American Phytopathological Society, St. Paul, Minn.

Csinos, A.S., Bell, D.K., Minton, N.A. & Wells, H.D. (1983). Evaluation of *Trichoderma* spp., fungicides, and chemical combinations for control of southern stem rot on peanuts. *Peanut Science*, **10**, 75–9.

Cullen, D. & Andrews, J.H. (1985). Benomyl-marked populations of *Chaetomium globosum*: survival on apple leaves with and without benomyl and antagonism to the apple scab pathogen, *Venturia inaequalis*. *Canadian Journal of Microbiology*, **31**, 251–5.

Davet, P. & Martin, C. (1985). Effets de traitements fongicides aux imides cycliques sur les populations de *Sclerotinia minor* dans le sol. *Phytopathologische Zeitschrift*, **112**, 7–16.

Davis, J.R., Fravel, D.R., Marois, J.J. & Sorensen, L.H. (1986). Effect of soil fumigation and seedpiece treatment with *Talaromyces flavus* on wilt incidence and yield, 1983. *Biological and Cultural Tests for Control of Plant Disease*, **1**, 18.

Driver, C.H. (1963). Further data on borax as a control of surface infection of slash pine stumps by *Fomes annosus*. *Plant Disease Reporter*, **47**, 1006–9.

Dubos, B., Jailloux, F. & Bulit, J. (1982). L'antagonisme microbien dans la lutte contre la pourriture grise de la vigne. *EPPO Bulletin*, **12**, 171–5.

Elad, Y., Chet, I. & Henis, Y. (1981). Biological control of *Rhizoctonia solani* in strawberry fields by *Trichoderma harzianum*. *Plant and Soil*, **60**, 245–54.

Elad, Y., Hadar, Y., Chet, I. & Henis, Y. (1982). Prevention with *Trichoderma harzianum* Rifai aggr., of reinfestation by *Sclerotium rolfsii* Sacc. and *Rhizoctonia solani* Kühn of soil fumigated with methyl bromide, and improvement of disease control in tomatoes and peanuts. *Crop Protection*, **1**, 199–211.

Elad, Y., Katan, J. & Chet, I. (1980). Physical, biological, and chemical control integrated for soilborne diseases in potatoes. *Phytopathology*, **70**, 418–22.

Fravel, D.R., Davis, J.R. & Sorensen, L.H. (1986). Effect of *Talaromyces flavus* and metham on *Verticillium* wilt incidence and potato yield, 1984–1985. *Biological and Cultural Tests for Control of Plant Disease*, **1**, 17.

Fravel, D.R., Marois, J.J., Dunn, M.T. & Papavizas, G.C. (1985). Compatibility of *Talaromyces flavus* with potato seedpiece fungicides. *Soil Biology and Biochemistry*, **17**, 163–6.

Garrett, S.D. (1965). Toward biological control of soil-borne plant pathogens. In: *Ecology of Soil-borne Plant Pathogens. Prelude to Biological Control*, eds K.F. Baker & W.C. Snyder, pp. 4–17. University of California Press, Berkeley, CA.

Garrett, S.D. (1970). *Pathogenic Root-infesting Fungi*. Cambridge University Press, Cambridge.

Hadar, Y., Chet, I. & Henis, Y. (1979). Biological control of *Rhizoctonia solani* damping-off with wheat bran culture of *Trichoderma harzianum*. *Phytopathology*, **69**, 64–8.

Henis, Y., Ghaffar, A. & Baker, R. (1978). Integrated control of *Rhizoctonia solani* damping-off of radish: effect of successive plantings, PCNB, and *Trichoderma harzianum* on pathogen and disease. *Phytopathology*, **68**, 900–7.

Henis, Y. & Papavizas, G.C. (1983). Factors affecting germinability and susceptibility to attack of sclerotia of *Sclerotium rolfsii* by *Trichoderma harzianum* in field soil. *Phytopathology*, **73**, 1469–74.

Katan, J. (1981). Solar heating (solarization) of soil for control of soilborne pests. *Annual Review of Phytopathology*, **19**, 211–36.

Katan, J. & Eshel, Y. (1973). Interactions between herbicides and plant pathogens. *Residue Review*, **45**, 145–77,

Katan, T., Dunn, M.T. & Papavizas, G.C. (1984). Genetics of fungicide resistance in *Talaromyces flavus*. *Canadian Journal of Microbiology*, **30**, 1079–87.

Kelley, W.D. & Rodriguez-Kabana, R. (1978). Effects of a pesticide on antagonistic soil fungi. *Highlights of Agricultural Research*, **25**, 10.

Klassen, W. (1981). The role of biological control in integrated pest management systems. In: *Biological Control in Crop Production*, ed. G.C. Papavizas, pp. 433–45. Allanheld & Osmun, Totowa, NJ.

Kraft, J.M. & Papavizas, G.C. (1983). Use of host resistance, *Trichoderma*, and fungicides to control soilborne diseases and increase seed yields of peas. *Plant Disease*, **67**, 1234–7.

Langerak, C.J. (1977). The role of antagonists in the chemical control of *Fusarium oxysporum* f. sp. *narcissi*. *Netherlands Journal of Plant Pathology*, **83** (Suppl. 1), 365–81.

Lantero, E., Bazzano, V. & Gullino, M.L. (1982). Tentativi di impiego della lotta integra nei confronti della *Botrytis cinerea* della vite. *Difesa delle Piante*, **5**, 11–20.

Lewis, J.A., Lumsden, R.D. Papavizas, G.C. & Kantzes, J.G. (1983). Integrated control of snap bean diseases caused by *Pythium* spp. and *Rhizoctonia solani*. *Plant Disease*, **67**, 1241–4.

Lewis, J.A. & Papavizas, G.C. (1980). Integrated control of *Rhizoctonia* fruit rot of cucumber. *Phytopathology*, **70**, 85–9.

Lewis, J.A. & Papavizas, G.C. (1984). Effect of the fumigant metham on *Trichoderma* spp. *Canadian Journal of Microbiology*, **30**, 739–45.

Locke, J.C., Marois, J.J. & Papavizas, G.C. (1985). Biological control of *Fusarium* wilt of greenhouse-grown chrysanthemums. *Plant Disease*, **69**, 167–9.

Lumsden, R.D., Lewis, J.A. & Papavizas, G.C. (1983). Effect of organic amendments on soilborne plant diseases and pathogen antagonists. In: *Environmentally Sound Agriculture*, ed. W. Lockeretz, pp. 51–70. Praeger Publishers, New York.

Maiti, D. & Sen, C. (1985). Integrated biocontrol of *Sclerotium rolfsii* with nitrogenous fertilizers and *Trichoderma harzianum*. *India Journal of Agricultural Sciences*, **55**, 464–8.

Marois, J.J. & Locke, J.C. (1985). Population dynamics of *Trichoderma viride* in steamed plant growth medium. *Phytopathology*, **75**, 115–18.

Marois, J.J., Johnston, S.A., Dunn, M.T. & Papavizas, G.C. (1982). Biological control of *Verticillium* wilt of eggplant in the field. *Plant Disease*, **66**, 1166–8.

Marois, J.J., Mitchell, D.J. & Sonoda, R.M. (1981). Biological control of *Fusarium* crown rot of tomato under field conditions. *Phytopathology*, **71**, 1257–60.

McKeen, C.D. (1975). Colloquium on integration of pesticide-induced and biological destruction of soil-borne pathogens: summary and synthesis. In: *Biology and Control of Soil-borne Plant Pathogens*, ed. G.W. Bruehl, pp. 203–6. American Phytopathological Society, St. Paul, Minn.

Mixon, A.C., Bell, D.K. & Wilson, D.M. (1984). Effect of chemical and biological agents on the incidence of *Aspergillus flavus* and aflatoxin contamination of peanut seed. *Phytopathology*, **74**, 1440–4.

Moubasher, A.H. (1963). Selective effects of fumigation with carbon disulphide on the soil fungus flora. *Transactions of the British Mycological Society*, **46**, 338–44.

Mughogho, L.K. (1968). The fungus flora of fumigated soils. *Transactions of the British Mycological Society*, **51**, 441–59.

Munnecke, D.E. (1984). Establishment of micro-organisms in fumigated avocado soil to attempt to prevent reinvasion of the soils by *Phytophthora cinnamomi*. *Transactions of the British Mycological Society*, **83**, 287–94.

Munnecke, D.E., Kolbezen, M.J., Wilbur, W.D. & Ohr, H.D. (1981). Interactions involved in controlling *Armillaria mellea*. *Plant Disease*, **65**, 384–9.

Papavizas, G.C. (1973). Status of applied biological control of soil-borne plant pathogens. *Soil Biology and Biochemistry*, **5**, 709–20.

Papavizas, G.C. (1981). Survival of *Trichoderma harzianum* in soil and in pea and bean rhizosphere. *Phytopathology*, **71**, 121–5.

Papavizas, G.C. (1985). *Trichoderma* and *Gliocladium*: biology, ecology, and potential for biocontrol. *Annual Review of Phytopathology*, **23**, 23–54.

Papavizas, G.C. (1987). Genetic manipulation to improve effectiveness of biocontrol fungi for plant diseases control. In: *Innovative Approaches to Plant Disease Control*, ed. I. Chet, pp. 193–212. Wiley, New York.

Papavizas, G.C., Dunn, M.T., Lewis, J.A. & Beagle-Ristaino, J. (1984). Liquid fermentation technology for experimental production of biocontrol fungi. *Phytopathology*, **74**, 1171–5.

Papavizas, G.C. & Lewis, J.A. (1979). Side-effects of pesticides on soil-borne plant pathogens. In: *Soil-borne Plant Pathogens*, eds B. Schippers & W. Gams, pp. 483–505. Academic Press, London.

Papavizas, G.C. & Lewis, J.A. (1981). Introduction and augmentation of microbial antagonists for the control of soilborne plant pathogens. In: *Biological Control in Crop Production*, ed. G.C. Papavizas, pp. 305–22. Allanheld & Osmun, Totowa, N.J.

Papavizas, G.C. & Lewis, J.A. (1983). Physiological and biocontrol characteristics of stable mutants of *Trichoderma viride* resistant to MBC fungicides. *Phytopathology*, **73**, 407–11.

Papavizas, G.C., Lewis, J.A. & Abd-El Moity, T.H. (1982). Evaluation of new biotypes of *Trichoderma harzianum* for tolerance to benomyl and enhanced bio-control capabilities. *Phytopathology*, **72**, 126–32.

Papavizas, G.C. & Lumsden, R.D. (1982). Improved medium for isolation of *Trichoderma* spp. from soil. *Plant Disease*, **66**, 1019–20.

Rasmussen-Dykes, C. & Brown, W.M. Jr (1982). Integrated control of *Pythium* blight on turf using metalaxyl and *Trichoderma hamatum*. *Phytopathology*, **72**, 976.

Richardson, L.T. (1973). Synergism between chloroneb and thiram applied to peas to control seed rot and damping-off by *Pythium ultimum*. *Plant Disease Reporter*, **57**, 3–6.

Rodriguez-Kabana, R., Backman, P.A., Karr, G.W. Jr & King, P.S. (1976a). Effects of the nematicide fensulfothion on soilborne pathogens. *Plant Disease Reporter*, **60**, 521–4.

Rodriguez-Kabana, R., Backman, P.A. & King, P.S. (1976b). Antifungal activity of the nematicide ethoprop. *Plant Disease Reporter*, **60**, 255–9.

Sewell, G.W.F. (1965). The effect of altered physical condition of soil on biological control. In: *Ecology of Soil-borne Plant Pathogens. Prelude to Biological Controls*, eds K.F. Baker & W.C. Snyder, pp. 479–94. University of California Press, Berkeley, CA.

Sivan, A., Elad, Y. & Chet, I. (1984). Biological control effects of a new isolate of *Trichoderma harzianum* on *Pythium aphanidermatum*. *Phytopathology*, **74**, 498–501.

Smith, R.J., Jr (1982). Integration of microbial herbicides with existing pest management programs. In: *Biological Control of Weeds with Plant Pathogens*, eds R. Charudattan & H.L. Walker, pp. 189–203. Wiley, New York.

Stern, V.M., Smith, R.F., van den Bosch R. & Hagen, K.S. (1959). The integrated control concept. *Hilgardia*, **29**, 81–101.

Strashnow, Y., Elad, Y., Sivan, A. & Chet, I. (1985). Integrated control of *Rhizoctonia solani* by methyl bromide and *Trichoderma harzianum*. *Plant Pathology*, **34**, 146–51.

Sundheim, L. & Amundsen, T. (1982). Fungicide tolerance in the hyperparasite *Ampelomyces quisqualis* and integrated control of cucumber powdery mildew. *Acta Agriculture Scandinavica*, **32**, 349–55.

Uytterbroeck, P., van Assche, C. & Spreuwers, J. (1979). Behaviour of *Trichoderma*-isolates after soil disinfestation and significance of antagonistic activity. In: *Proceedings of the International Symposium of the IOBC/WPRS* on Integrated Control in Agriculture and Forestry, Vienna, 8–12 October 1979, eds. K. Russ and H. Berger, pp. 429–32.
IOBP = International Organisation for Biological Control
WPRS = West Palearctic Regional Section

van Dommelen, L. & Bollen, G. J. (1973). Antagonism between benomyl-resistant fungi on cyclamen sprayed with benomyl. *Acta Botanica Neerlandica*, **22**, 169–70.

Wainwright, M. & Pugh, G.J.F. (1975). Effect of fungicides on the numbers of microorganisms and frequency of cellulolytic fungi in soils. *Plant and Soil*, **43**, 561–72.

Yelton, M.M., Hamer, J.E. & Timberlake, W.E. (1984). Transformation of *Aspergillus nidulans* by using a trpC plasmid. *Proceedings of the National Academy of Sciences USA*, **81**, 1470–4.

Index

241, 248
Anticarsia gemmatalis, 48, 49
ants, 10
Aphanocladium, 171
Aphanomyces, 172
aphicides, 30
Aphidius matricariae, 25
Aphidoletes aphidomyza, 25
aphids, control with *Verticillium lecanii*,
  14, 22–31, 39–42, 69, 79, 177,
  213–15, 220–1, 225; resistance to
  pesticides, 20
Aphis fabae, 28, 42; *gossypii*, 23–4,
  28, 31, 52
apomicts, 118
apple, 13, *see also Malus*; aphids, 42;
  Bramley's Seedling, 150–1; dry eye
  rot, 145; scab, 5, 147, 248
application of biocontrol agents,
  methods, against plant pathogens,
  130, 145–7, 149–50, 155, 177,
  delivery systems, *175–81*; against
  insects, 30–2, 42, 44–6, 50–2, 66,
  156, 220; against weeds, 90–5, 98,
  102, 104, 111; with pesticides or
  chemicals, 11, 31–2, *236–49*
appressorium, 39, 153–5, 170, 192, 214
apricot, 151; die back, 244
Armadillium, 23
Armillaria mellea, 5, 237, 238
Arthrobotrys, 189; *amerospora*, 198;
  dactyloides, 4, 194; *oligospora*, 4,
  201; *robusta*, 197
arthropods, 7, 13, 22–3, 26
artichoke, 116
Aschersonia, 38; *aleyrodis*, 24, 42–3, 62
Ascochyta caulina, 96; *pteridis*, 6, 12, 96
Ascomycetes, 5, 7, 9, 128, 132, 242
ascospores, 242–3, 248
Aspergillus, 73–4, 765, 132, 203, 238,
  240, 242; *awamori*, 74–5; *flavus*, 75,
  246; *fumigatus*, 226; *glaucus*, 127;
  nidulans, 222, 226–8, 248; *nidulans* x
  fumigatus, 226; *oryzae*, 75–6;
  parasiticus, 75
Aspidiotus, 23
Astacus astacus, 218
asulam, 6
attapulgite, 72
aubergine, 23
augmentative weed control, 2, *91*;

strategy, 91
Aulocorthum, 23
Aureobasidium pullulans, 147, 155
autolysis, 76
auxotrophs, 219, 222–6; markers, 214;
  mutants, 219, 226; recipient protoplasts, 227; parental protoplasts, 227

bacilli, 149–50
Bacillus, 10, 148, 149; *cereus*, 149;
  cereus subsp. *mycoides*, 150–1;
  pumilus, 149; *subtilis*, 150–1;
  thuringiensis, vii, 7, 10, 23, 32, 150,
  229
bacilysin, 150
bacteria, 2, 3, 5, 7, 10, 15, 20, 73,
  75, 90, 95, 142–3, 146–7, *148–52*,
  153–5, 202, 229
baculovirus, 46
bags, fermentation, 75
bait, 46, 68–9
banana, 154
bark, 180
barley, 72
barnyard grass, 97
basidiocarp, 178
Basidiomycetes, 7–9, 128
basidiospores, 178
baskets, fermentation, 76
Bathurst burr, 97
beans, 23, 42, 243; dry, 150; French,
  154; rust of, 150–1; snap, 150, 224
beauveracin, 217
Beauveria, 7, 38, 62, 68, 78, *bassiana*,
  24, 38, 39, 40, *43–4*, 48, 50–4, 62,
  66–7, 69, 78–9, 212–17; *brongnartii*,
  41, 44–5, 62, 79, 213, 215–16, 218,
  (*tenella*), 67–8, 212
beer mash, 68
beetles; *see also* Coleoptera; Colorado
  potato, 7, 43–4, 69, 214, *see also*
  Leptinotarsa decemlineata; may, 44;
  rhinoceros, 45, 214; scarabid, 214
beetroot, 148–9, 153
begonia, 102
benlate, *see* benomyl
benodanil, 32
benomyl, 14, 31–2, 101, 147, 224, 227,
  240, 245–8
bentazon, 101
Bermuda grass stunt mite, 69

Xanthium spinosum, 97; *strumarium*,
   96

yeasts, 7, 26, 54, 142–8, 150, 153–5,
   227; basidiomycetous, 145

Zoopagales, 4
Zoophthora, 38; *radicans*, 38, 41–2

zoosporangia, 192, 240
zoospores, 63–4, 132, 191–2, 196
zoosporogenesis, 64
Zygina pallidifrons, 23
Zyginidia pullula, 25
Zygomycetes, 4, 62–3, *64–5*
zygospores, 132
zymogenous fungi, 195